Orchestrating Success

Orchestrating Success

Improve Control of the Business with Sales & Operations Planning

Richard C. Ling
and
Walter E. Goddard

John Wiley & Sons, Inc.

New York · Chichester · Brisbane · Toronto · Singapore

ISBN: 0-471-13227-6

Printed in the United States of America

10 9 8

Contents

Preface

Every manager seeks control—the ability to make the right thing happen. Surprises fill the absence of good controls. Even pleasant surprises are problems, as they indicate that things did not happen as expected.

Sales & Operations Planning provides the general manager and his staff a means of gaining more control. With it, they can operate their business more effectively: set attainable objectives, see consequences, evaluate alternatives, communicate approved plans, measure performance, and achieve predicted results.

Faster and more economical responses to the marketplace are the major benefits. There are two equally impressive by-products: the satisfaction and confidence of everybody within the organization. Valid plans and controlled changes make everyone's job more productive and fulfilling.

Input from the business plan and conversion into detailed sales and production plans are the activities that surround the Sales & Operations Planning process. None of these work well individually. Although this book purposely emphasizes the Sales & Operations Planning function, it is not a stand-alone activity. When combined with the others, however, it can make for the difference between controlled actions and chaos.

Sales & Operations Planning works well because it is so logical and straightforward. We expect much of our advice to sound familiar and simple. We have not created a new activity, but rather are presenting a proven one in a structured manner. Yet, extraordinary results occur whenever common sense is organized properly.

This book is designed to help those managers who want to understand what Sales & Operations Planning is all about and to share with all managers how to use it successfully. It will be a compliment to us if you easily grasp the contents, and it will be of great value to your company when you quickly apply them.

We are proud to author the first text devoted to Sales & Operations Planning. It represents, however, a collective effort, which incorporates the experience of our colleagues and client companies. We hope they will be pleased with our book.

Richard C. Ling
Winston-Salem, NC
March 1988

Walter E. Goddard
Sunapee, NH
March 1988

Acknowledgments

First, none of this would be possible without our mentor and friend, the late Oliver Wight. He provided the foundations that enabled us to develop Sales & Operations Planning as well as the inspiration to share our knowledge. It is on our combined years of teaching and consulting that this book is based.

A tip of the hat goes to our associates Chris Gray, Darryl Landvater, John Sari, and Peter Skurla for their invaluable help in reviewing and adding to the many drafts of this book.

Special thanks to our publisher, Dana Scannell, without whom we never would have hit a schedule, and to Steve Bennett, a professional writer who made each of our ramblings more intelligent and consistent.

We would also like to thank Roger Brooks, John Dougherty, George Palmatier, Mike Salmon, John Schorr, Bob Stahl, Al Stevens, and Tom Wallace for their contributions as co-instructors and associates. This group of people has greatly influenced our thinking and the evolution of our subject over the years.

In addition, we would like to thank a group of people who shared with us their many years of experience using Sales & Operations Planning, and those who used a rough draft of this book as a guide to test its validity as a "how-to" document. These people include: C. Dow

Caldwell, Abbott Laboratories; Thomas Connelly, Hardinge Brothers; Robert Egan, Hardinge Brothers; Charles Glinsky, Jamesbury Corporation; Allison Gray, Kingston-Warren; Peter W. Johnson, ICI Americas; Richard W. Kraber, Garden America; Kate May, Molex; and John Trotter, Rexnord.

Last, but certainly not least, thanks go to our wives, Natalie Ling and Ellie Goddard, for their unending patience and encouragement. It was their support that finally forced us to stop talking about our great plans and to start writing.

Orchestrating
Success

Chapter 1

Working in Concert

There are two golden rules for an orchestra:
start together and finish together.
(Sir Thomas Beecham, British conductor)

Consider the following scenarios. Three orchestras begin playing Beethoven's Fifth Symphony. The first orchestra works like a well-oiled machine; all sections mesh in perfect harmony as they take their cues from the conductor. The result is magnificent music that sweeps the audience away. "Truly inspired," one reviewer dreamily writes in his column for the *Tribune*.

In a neighboring city, the second orchestra isn't faring as well; the woodwinds and strings are not quite tuned to each other. Although the discord is slight, it is nonetheless annoying. The poor conductor wishes he could stop the piece, retune, and start again, but unfortunately, this isn't a rehearsal. "How did this happen?" he asks himself. In the front row, the music columnist for the *Daily Star* is pondering the same question.

And in another nearby city, the third orchestra is having a terrible night. The woodwinds, brass, and strings are fighting each other as they go their own ways. The conductor, caught in a cross fire of noises, winces as people in the audience begin leaving the concert hall. The columnist from *The Bugle* considers switching from the arts section to the comics as he watches three members from the brass section sneak off the stage with their hands covering their ears.

3

Aren't these three orchestras like the manufacturing companies of America? Some companies, like the first orchestra, are well synchronized; their sales, marketing, manufacturing, engineering, and finance departments work together and share information so that they can economically match production to the demands of the marketplace as quickly as possible. Such companies achieve stunning customer service levels and enjoy handsome growth and profits in good and bad economic times alike. They also lead their industries by commanding and creating markets, and setting the standards for worldwide competition.

Many companies are like the second orchestra—they experience a degree of discord among departments that renders them less competitive and profitable than they ought to be. Their sales and marketing departments have misperceptions about the capabilities of the manufacturing and engineering departments, while their manufacturing and engineering departments distrust the forecasts and sales plans created by sales and marketing. Such problems may be accentuated by a language barrier. Sales and marketing usually talk dollars, while manufacturing and engineering need to plan in terms of capacity and man hours. If not handled properly, this discrepancy can cause communication problems, which in turn can lead to compromised performance.

Finally, there are manufacturing companies that, like our third orchestra, are literally torn apart by internal wrangling. Their various departments not only lack communication, but actually consider themselves to be in direct competition with each other. When this happens, the company performs miserably. Customer service suffers as shortages rise and promised delivery dates are missed. Hoped-for profits evaporate as the company pays premium charges for materials and labor to compensate for production delays. Upset customers begin going elsewhere. The result? The management team might soon find itself playing a sad song.

The fact is, it doesn't take a maestro to convince all departments to work in concert. But it does require a competent conductor who can keep everyone playing the same score on the same page. Any good general manager can harmonize his company through a process called

"Sales & Operations Planning,"[1] in which he meets with his top managers on a regular and frequent basis to update the plans for all departments. The plans take into account projections made by the sales and
marketing departments, the resources available from manufacturing,
engineering, purchasing, and finance, and are directed toward hitting
the company's objectives. Sales & Operations Planning is done on an
"aggregate" or "family" level, and covers a sufficient span of time
to make sure that the necessary resources will be available. The approved aggregate plans drive the individual departmental detail plans.
Each month—or more frequently if the market conditions are volatile—the representatives meet again to determine whether the overall
company plan is on course, and to adjust for changes in the marketplace and changes or problems within the company.

The term "general manager" is used throughout this book, and we
need to define its meaning since titles vary so greatly from company
to company. The general manager is the person within your company
who manages four key functions: marketing, manufacturing, engineering, and finance. These four areas are critical because people within
them will recommend changes to existing plans and will be responsible
for executing the approved plans. The executive to whom they report
is the final decision-maker in the planning and execution process. In
many companies this executive might bear another title, such as "president," "chief executive officer," "chief operating officer," or "executive vice president."

WHAT'S IN IT FOR YOU?

Every company usually has some kind of regular monthly staff meeting in which the financial numbers are reviewed and some marketing
and manufacturing issues are discussed. There are a number of major
differences between these sessions and a Sales & Operations Planning
meeting. Without thought and structure, the regular monthly meetings

[1] Sales & Operations Planning (S&OP) should not be confused with Standard Operating Procedure (SOP). But we believe that S&OP should become a routine part of SOP in *all* manufacturing companies.

tend to deal with near-term responses to problems and opportunities, and often lack sufficient information to determine the consequences and evaluate alternatives.

Sales & Operations Planning offers the manufacturing company many advantages over the weekly staff or peer group management meeting. These differences translate into four major benefits to any manufacturing company. First of all, Sales & Operations Planning provides the necessary link between the company's business plan and the operations of each department. Once an effective Sales & Operations Planning process is in place it will ensure that the operating plans are in lockstep with the business plan, or make it evident where sales and production have deviated from the business plan. This is necessary so that course corrections can be made and the company can better achieve management's objectives. This validation aspect is also critical, because it means the difference between having your fingers crossed and having confidence, the difference between saying, "We'll try our best" and "We can do it." In short, validation makes performances match promises—it's a contradiction to say that you have good business plans if you're uncertain as to whether they can be carried out.

Another form of validation centers on units of measure. The business plan always talks dollars, while the operating plan deals in other units of measure, such as pieces, tons, yards, standard hours, etc. Each has its purpose, as it represents how the company and its members need to assess the plans. Therefore, both dollars and other units of measure are needed. Dealing in different units of measure, though, creates the risk that they may inadvertently get out of step with each other. The unit-of-measure validation is needed to prevent such mismatches through a two-way conversion between the dollars and the business plan and the other units of measure used in the Sales & Operations plan.

The second major benefit of Sales & Operations Planning is that it provides a means for orchestrating all departments, as it communicates the plans both horizontally and vertically. The horizontal communication occurs across all of the company functions. This ensures that each functional area knows the company's objectives so that each can initiate what it needs to do to carry out the objectives. Without a common plan, the company will expend far more effort and realize far fewer results.

The vertical communication occurs within each department, through either a "top-down" or a "bottom-up" planning process that results in the aggregate and detailed plans being in sync. The plans spell out in aggregate what each department must do. Only when people in a department know they can't support the plan should they speak up, identifying the cause of the problem and recommending the best possible solution. This must not occur only within the individual departments; each department must communicate its ability to carry out the plan at the Sales & Operations Planning meeting.

The third benefit of Sales & Operations Planning is that it yields a realistic plan capable of achieving the company's objectives. This is important, because the most brilliant plan in the world will be of little value if it does not produce the desired results. The Sales & Operations Planning process provides a "reality" check, as it requires input and sign-off from all departments. Such an approach has another important benefit; since everyone participates in the decision, a genuine sense of teamwork and ownership in the plan arises. With teamwork, the need for defensive maneuvers and finger-pointing fades away and morale improves. A related benefit is that it provides a medium through which each department can better understand what the other departments can or can't do. Only when that understanding exists can conflicts get resolved properly through the shared goal of meeting the company's objectives. The best possible customer service at the lowest possible cost with optimum profits occurs when everyone pulls together in the same direction.

Finally, Sales & Operations Planning eliminates what Manufacturing Resource Planning (MRP II) pioneer Oliver Wight called "hidden decisions." No department can exist in a vacuum—many decisions that one department makes ultimately affect another. Sometimes a domino or chain reaction is started, and the decision of one department may affect the company as a whole. If the decisions that are made conflict with the real capabilities of another department, the result will be "random scatter," in which the dominoes fall without any specific organization. In contrast, Sales & Operations Planning shifts the company from managing by "default" (i.e., by hidden decisions and hidden agendas) to implementing an explicit and visible set of plans and decisions arrived at by people who share their information and work toward a common goal. It is a way of initiating a "positive domino effect," in which good information and communication breed good decisions, so that the whole company benefits. The chain reaction leads to directed action rather than random scatter.

Like any process that offers enormous benefits, Sales & Operations Planning requires work. The general manager must be capable of getting his staff to arrive at a master game plan. Additionally, there must be a process that breaks down the aggregate plan into detailed plans. MRP II and Just-in-Time are the accepted means of accomplishing this task. MRP II is an organized approach to planning and controlling all of the resources required to operate a manufacturing business: materials, manpower, equipment, tooling, space, and money. Just-in-Time is an organized company-wide effort aimed at changing the environment by eliminating non-value-adding activities, simplifying those that cannot be eliminated, and excelling all value-adding activities. For a more detailed explanation of MRP II and Just-in-Time, see Appendices B and C.

Once the Sales & Operations Planning process is started, it must be continued every month to achieve optimal results. You must ensure that all the necessary plans get communicated and that people can arrive at a master game plan.

Whatever the effort, though, the rewards will be more than worth it. Most companies realize immediate benefits from the Sales & Operations Planning process because it forces them to improve their interdepartmental communications. Such improvements can only lead to better-than-before performance. If your company is implementing or using MRP II and Just-in-Time, an effective Sales & Operations process is required to achieve the maximum potential results. In any case, all companies eventually discover that Sales & Operations Planning gives them a new measure of control over their business.

In the following pages you'll learn about the essential concepts necessary for carrying out Sales & Operations Planning. They are intended to give the general manager and his staff the understanding required to implement Sales & Operations Planning or improve their present process. Chapter 2 explains the basic Sales & Operations Planning process. It also describes how various departments should be represented at the Sales & Operations Planning meetings, and points out some pitfalls that may delay the process. Chapter 3 discusses the general manager's role in the Sales & Operations Planning process, and Chapter 4 discusses what each department needs to prepare so that it can participate in a Sales & Operations Planning meeting. Chapter 5 covers data gathering and presentation. Chapter 6 focuses on the meeting itself, walking you through each step that a planning team must take during a typical session. The seventh and final chapter takes a step back and describes the process of implementing a Sales & Operations Planning process in your company. It covers who should spearhead and oversee the effort, and what steps need to be taken to make Sales & Operations Planning a reality.

We cannot guarantee that carrying out Sales & Operations Planning will bring you an instant rise in profits or productivity. We do promise, however, that by adopting the principles and ideas you're about to read, your company will function as a more harmonious whole. And that should be music to any general manager's ears.

SUMMARY

- It takes a good conductor/general manager to get all of the departments working together.
- Sales & Operations Planning deals with families of products and requires a sufficient planning horizon as well as monthly reviews.
- Sales & Operations Planning links the business plan with the operations of each department.
- Sales & Operations Planning communicates plans horizontally, across company functions, and vertically, by department.
- The Sales & Operations Planning process yields realistic plans.
- Sales & Operations Planning eliminates surprises and hidden decisions.

Chapter 2
Synchronizing with the Marketplace

The conductor is there . . . first of all for the oversimplified reason of just being the traffic cop, making sure everyone is playing at the same speed and the same volume.

(André Previn, American conductor)

Sales & Operations Planning is a dynamic process in which the company operating plan is updated on a regular monthly or more frequent basis. Here's a capsule description of the process. It starts with the sales and marketing departments comparing actual demand to the sales plan, assessing the marketplace potential and projecting future demand. The updated demand plan is then communicated to the manufacturing, engineering, and finance departments, which offer ways to support it. Any difficulties in supporting the sales plan are worked out, or the sales plans are altered in a process that concludes with a formal meeting chaired by the general manager. The final result is a set of ''marching orders'' for all departments that extends through the current fiscal year and as far beyond that as is necessary to effectively plan resources. Most important, an updated operation plan is being set to satisfy the current market, and the consequences of taking various actions are known ahead of time, minimizing costly and disruptive surprises.

New Englanders are famous for their frugal use of words to make a succinct point. Two old Yankee proverbs relate well to Sales & Operations Planning. The first, ''He sees best who sees the consequences,'' means that predicting consequences prior to approving an action plan

11

Figure 2.1

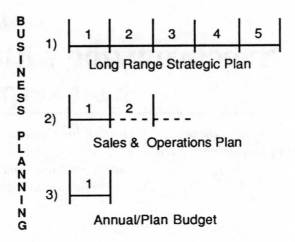

is an important ingredient in good decision-making. The second, "Better to be ready and not go than to go and not be ready," nicely states the value of contingency planning and "what-if" analysis.

The essence of good managing is contained in these proverbs. Every manager should be thinking ahead and evaluating options and alternatives before they are needed. Such thinking does not indicate doubt that the selected plan is the best one, nor does it reflect lack of resolve to make the plan happen. Rather, it characterizes a prudent manager, one who expects changes and is therefore better prepared to handle them.

Another contribution of Sales & Operations Planning is that it enables a company to fine-tune its long-range strategic plan and annual business plan. To understand how this fine-tuning process works, let's consider the overall business planning process shown in Figure 2.1.

At the top level of the process is the long-range strategic plan, which is reviewed on an annual basis. The strategy must answer such vital questions as:

• What is our business (products and services)?
• To whom do we sell (markets, customers)?

- What resources are required (people/skills, technology, plant and equipment, distribution, etc.)?
- What is the measure by which we compete (quality, delivery, price, service)?
- What is our financial strategy (profit, growth, ROA, ROI)?
- Any additional strategies (make or buy, market share, flexibility)?

The middle level of the process shown in Figure 2.1, the ongoing operating plan, entails developing specific goals and "how tos" through the Sales & Operations Planning process. At the bottom level we encounter the annual business plan, which is used for financial planning and measurement purposes, as well as for communicating with the financial community.

Whereas the long-range strategic and annual business plans are updated on a yearly basis, the ongoing operations plan is continually revised through the Sales & Operations Planning process. This feature, coupled with the fact that the ongoing operating plan often extends well beyond the annual business plan, enables companies to develop business plans that are not only consistent with long-range strategic goals, but realistic in terms of the marketplace.

This approach is radically different from the all-too-traditional view of the strategic and business plans as carved in stone. The drawback to such a one-shot approach is that market conditions and your capabilities constantly change, and static plans can quickly become outdated. Nevertheless, many businesses still manage to the "original numbers," sticking to the financial targets regardless of what is actually happening in the outside world and inside the company. As a result, the business battlefield is strewn with the bodies of companies that "hung in there till the end of the year" with unrealistic expectations that somehow the plan would be met. Other companies fail to make "opportunity decisions" when demand shifts to give them a chance for greater market share. In either case, performance is worse than it would have been if the companies had revised their plans and appropriately responded to the ebb and flow of the marketplace.

Sales & Operations Planning offers just such a means for achieving and updating the business plan. As C. Dow Caldwell, director of materials management at Abbott Laboratories, told us, "Looking ahead to plan and avoid problems is vital. In business, that starts with strategic planning. But to be effective, the implementation of the strategic plan must be an ongoing process to adjust to change. Also, it is necessary to review plans in the aggregate to keep management's viewpoint in focus. That's what Sales & Operations Planning is all about.''

Because it takes place on a regular and frequent basis, Sales & Operations Planning offers a window through which changes in the marketplace and required company actions become quite visible. This enables the company to adjust its production levels to avoid excess inventory and backlog, to gear up its operation so that it can seize new opportunities for greater market share, or to adjust its plans for changes in product mix. In turn, the impact of these changes can be reconciled with the business plan.

Sales & Operations Planning offers another key benefit to companies striving for the competitive edge: it enables them to operate Manufacturing Resource Planning (MRP II) at its full potential. Many people assume that MRP II is primarily a computer-supported approach to scheduling activities within the factory. While hardware, software, and scheduling are critical aspects of MRP II, the best computer in the world won't improve your competitive posture unless it's carrying out the plans approved by the general manager. In fact, without Sales & Operations Planning, MRP II isn't MRP II—it's little more than a middle-management operating system, and will yield only a partial payback.

In contrast, when Sales & Operations Planning is the driver in an MRP II system, the results are likely to be Class A performance, which in turn means better customer service, more reliable performance, reduced costs, and greater profits. How the various plans function at the aggregate and detail level within the context of MRP II is shown in Figure 2.2. While it is beyond the scope of this book to discuss all aspects of MRP II in depth, Appendix E offers a source for excellent books on the subject.

Figure 2.2

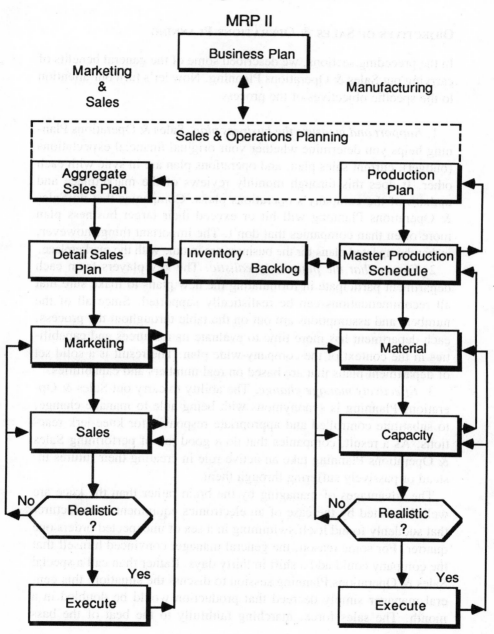

OBJECTIVES OF SALES & OPERATIONS PLANNING

In the preceding sections, we described some of the general benefits of carrying out Sales & Operations Planning. Now let's turn our attention to the specific objectives of the process:

1. *Support and measure the business plan.* Sales & Operations Planning helps you determine whether your original financial expectations (budget), current sales plan, and operations plan are in sync with each other. It does this through monthly reviews of the marketplace and updates of the company's operations plans. Companies that do Sales & Operations Planning will hit or exceed their target business plan more often than companies that don't. The important thing, however, is that the updated plans for the business are in tune with the marketplace.

2. *Ensure that the plans are realistic.* The key players from each department participate in formulating the new plans to make sure that all recommendations can be realistically supported. Since all of the numbers and assumptions are out on the table throughout the process, each department has more time to evaluate its resources and capabilities in the context of the company-wide plan. The result is a solid set of department plans that are based on real numbers and capabilities.

3. *Effectively manage change.* The ability to carry out Sales & Operations Planning is synonymous with being able to manage change, to substitute controlled and appropriate responses for knee-jerk reactions. As a result, companies that do a good job of performing Sales & Operations Planning take an active role in creating their futures instead of passively suffering through them.

The advantages of managing by the brain rather than the knee are well illustrated by the case of an electronics equipment manufacturer that suddenly found itself swimming in a sea of unexpected orders one quarter. For some reason, the general manager convinced himself that the company could add a shift in thirty days. Rather than call a special Sales & Operations Planning session to discuss the situation, this general manager simply decreed that production would be doubled in a month. The sales force, marching faithfully to the beat of the bass drummer, confidently began promising orders based on product being available because of the new shift. But the hiring and training just didn't happen in thirty days. As a result, the company had to go to significant overtime in the hopes of solving the problem. And despite the overtime, promised ship dates were missed, customer service declined, and costs were higher and profits lower.

Equally disastrous, as production struggled the company began accumulating raw materials and mismatched components, which merely tied up cash and misused capacity. The net result? Significant opportunities for sales and profits were lost. And it took six months to dig out from under the one-minute, one-man decision and restore reliable service to customers. As this company painfully learned, when the conductor ups the tempo he had better be sure that all players can keep up with the new speed.

4. *Better manage finished goods inventory and/or backlog to support customer service.* Maintaining the right level of finished goods inventory for make-to-stock families is essential for good "off-the-shelf" customer service. Operating at too high an inventory level results in extra costs, while operating at too low a level creates too many back orders. In a similar manner, controlling backlogs for make-to-order products is also essential for good customer service. If actual backlog becomes too large, delivery times stretch out, which eventually will cause customers to go elsewhere. By contrast, insufficient backlog can incur extra operating costs.

The general manager and his staff have the responsibility for establishing targets for what levels of inventory and backlog they believe are necessary to remain competitive. If the general manager and his staff don't establish targets by families, then by default it will be done by people at lower levels as they make individual decisions. Seldom will the sum of these detailed decisions add up to an aggregate plan that would represent what the general manager would have done himself. The linkage of detailed decisions to aggregate is a vital part of controlling the business, ensuring that each is supporting the other.

Maintaining the desired levels of finished goods inventory and backlog is an ongoing challenge for two reasons. First, it may be difficult to gain consensus on what the future targets should be, and second, it will usually be hard to hit the targets economically. Nevertheless, the process of reviewing the targets, discussing the consequences of changing them, and finally approving them is an effective means to meet the challenge.

5. *Measure performance.* Sales & Operations Planning incorporates performance measurements to identify whenever actual performance has deviated significantly from the plan. The two main purposes of this are to separate those activities that are in control from those that aren't, and to quickly bring the out-of-control situation to the surface so that an evaluation can be made and, if necessary, corrective action taken.

Measuring performance against plans is only productive when the plans are valid. This is a very important contribution that Sales & Operations Planning makes to a company. Whenever the targets are challenging but attainable, managers are willing to be held accountable for their performance.

6. *Build teamwork.* A key element of Sales & Operations Planning is that it gives each department an opportunity to participate in the overall planning process. Each executive brings his experience and skills, which add insights to the matter of making changes to the current plans. These same talents can respond to proposed changes in terms of consequences and alternatives. This not only ensures that the general manager is receiving the best possible advice before approving the new plans, it also demonstrates that each staff member is an important and valued part of the team. Thus the process not only provides a means of updating the operating plan to bring it into step with changes in the marketplace, it also instills a spirit of teamwork and a shared set of goals in achieving the new company plans. The result of such teamwork, of course, is a better-performing company.

PREREQUISITES FOR PERFORMING SALES & OPERATIONS PLANNING

As we stated earlier, *any* company can effectively use Sales & Operations Planning to improve its performance and become more competitive. Five prerequisites, however, need to be in place:

1. Each department must gain an understanding of the Sales & Operations Planning process.

2. The company must commit the time and resources to the process.
3. The company must define product groupings or families.
4. The company must establish an adequate planning horizon.
5. The company must establish and manage time fences.

These topics are discussed in the following sections.

1. Understanding Sales & Operations Planning

For Sales & Operations Planning to be effective, there can be no "black boxes" in the process; all participants must understand how it works and what it is designed to achieve. When people understand that sharing information does not mean giving up control and they see that the exchange actually leads to gaining control, they will be more willing to work in concert with their fellow departments toward the larger objectives of the company.

2. Commitment and People

Once a company embarks on the Sales & Operations Planning process, it is really making a lifetime commitment. Each month, or possibly more frequently, the decision-makers from all departments, along with the general manager, must review and update the company's sales, production, engineering, and financial plans. This "core team" is made up of the top executives from sales, marketing, manufacturing, engineering, finance, and human resources.

Figure 2.3 suggests the personnel who should participate in the Sales & Operations Planning sessions. Note the general manager at the top of the list; it is mandatory that he attend all Sales & Operations Planning sessions if the process is to be successful. Once the general manager commits fully to the process, the other key people will set aside time to meet regularly, regardless of their other activities. In business, as in shepherding, the flock will follow the leader's actions.

Once the process becomes part of the routine of the company, people will no longer treat Sales & Operations Planning as a matter of setting aside extra time to attend the meeting, but rather will discover that the meetings are the best use of their time, and to discontinue them

Figure 2.3
Sales & Operations Planning
Typical Participants

MANDATORY PARTICIPANTS	POTENTIAL PARTICIPANTS
General Manager	
Sales	
Department Manager	Customer Service Manager
	Distribution Manager
	Service Parts Manager
	Demand Manager
Marketing	
Department Manager	Chief Forecaster
	Product Manager(s)
Manufacturing	
Department Manager	Manufacturing Manager
Materials Manager	Master Scheduler
	Purchasing Manager
	Quality Assurance Manager
Engineering	
Department Manager	Drafting Manager
	Engineering Scheduler
	Manager, Design Engineering
Finance	
Department Manager	Budget Manager
	Cost Accounting Manager
Human Resources	
Department Manager	
Programs/Special Projects	
Appropriate managers	

would be a guarantee of far more effort on everyone's part while generating fewer results.

Initially, care must be taken to ensure that dates are on everyone's calendar, and that sufficient time is allocated to do justice to the issues raised in the meetings. Of course, in the event of an emergency that would prevent a member of the core team from attending, a substitute from that department must attend. This person must have the authority to represent the needs of his department, and be empowered to make the important decisions.

Many companies do not operate the Sales & Operations Planning meeting with only the core team in attendance. Rather, they have found it useful to have a number of other people available on a "need for" basis. For example, take a situation that on the surface seems straightforward: the rate of output for one family needs to be increased, while at the same future point in time, another family needs to be decreased by the same amount. Is this a straight trade? Are the skills transferable? Are the equipment and tooling common? Are there any problems with procuring raw materials? In short, is it a one-for-one exchange?

If the core group has general knowledge and recognizes that there may be a number of aspects of which they aren't aware, but which could be critical to the execution of the plan, they would either have to make a risky decision or postpone it in order to do further analysis. An alternative would be to have the appropriate people with the expertise give their more detailed advice and recommendations. There are times when the need for the greater amount of detail is known in advance, and in these cases the appropriate people can be invited to attend. Where this happens frequently and/or unexpectedly, it may be more appropriate to have the additional people attend regularly. Where the meetings are lengthy, and some of these extra people are not involved in all aspects, it may be better to have them on call or ask them to sit in for only those sections of the meeting where their experience is pertinent. The point is to get the right mix of power and knowledge into the planning and meeting process so that the best possible decisions can be made and executed.

This is particularly crucial when managing through a period of significant change or difficulty, when decisions need to be made about allocating scarce capacity to products, or changes in production must be made within a short lead time. In these circumstances, product managers, factory management, and master schedulers might be the right people to supply information about which course to follow when the alternatives and decisions are not easy.

3. Defining Families

Sales & Operations Planning is carried out at the aggregate level. By "aggregate" we mean product groupings or families rather than individual products or items. Why manage at the aggregate level? Because it's just not practical for top management to juggle every item that the company manufactures. The idea is to get effective input and control from management. This comes about by managing families, not items, and managing rates, not specific work orders.

Managing in the aggregate means grouping products into logical families. This may be straightforward if all parties agree what the families should be. Very often, however, sales and marketing view things in aggregate differently than manufacturing does. Sales and marketing naturally look at their products the way customers look at them, from a standpoint of function and applications. Manufacturing, in contrast, tends to look at products in terms of processes.

At first blush, such differences would seem to be an explosive mismatch. And many companies do, in fact, wind up getting ensnared in nonproductive debates about "who's right." This is precisely what happened at a major flatware manufacturer, which offered a variety of patterns for each of its products. While the marketing department was concerned with styles, the manufacturing department was concerned with machining capacity. Which department was right? Both. The various patterns had tremendous differences in the amount of required capacity. For instance, the more expensive patterns required a great deal more finishing and handwork, and hence more labor. As a result, manufacturing families could almost be cut along value lines—the more valuable the product, the more costly the manufacturing process. Marketing, by contrast, viewed product lines in terms of style, which did not necessarily relate to manufacturing's family definition.

Although both departments had valid points of view, this company organized its families according to sales and marketing's perspective. Manufacturing then made conversions from units to capacity, which is how it should be. After all, customers buy products because of what they do, not how they're made—the marketplace typically doesn't relate to the manufacturing process. As a result, sales and marketing should establish families that make sense to them, and the means should be in place for converting the plans by family into meaningful terms for other departments.

When departments have differing views of families, a conversion table will be necessary. Such a table can be established by analyzing past history to determine what impact each marketing family has on the manufacturing families. This approach can be very effective in planning manufacturing resources. The conversion factors need to be monitored, though, because there will be changes over time due to events in the marketplace and/or the installation of new equipment. Many companies utilize the rough-cut capacity planning capabilities in their standard software to perform the conversions that we'll discuss in Chapter 4.

Ideally, the families that you select should meet two criteria: size and meaningfulness. The larger the family the better, for two reasons. First, it's less work, and second, forecasting is always more accurate for broader groupings. For the families to be "meaningful," the general manager and his staff should be able to relate to these families in a manner that enables them to respond to requests to change output up or down.

These two characteristics are often in conflict. For example, a company that manufactures office furniture consisting of chairs, file cabinets, and desks would find that combining all three into one family would result in too few, and three may not be enough. Simply stating that sales are to increase by 25 percent in total would not enable manufacturing to respond with confidence as to whether they could quickly handle the increase. Rather, manufacturing would need to know whether each of the product families was to increase evenly, recognizing that a 25 percent total increase in business could originate from a 40 percent change in one product line and flat sales in the other two. In this case, neither the skills nor equipment necessary to manufacture each product line are transferable. Thus, at the least manufacturing would need to analyze the request to change output rates as a function of the three major processes used to create the products. Moreover, if the company made metal desks and wooden desks, the decision about increasing output would very likely require a further subdivision of families for the same reasons: lack of interchangeability of operators and equipment.

As an aid to resolving whether you have defined families properly, we recommend the "frown test." Create the minimal number of families, and check the reaction from manufacturing, engineering, and marketing. Can they comfortably deal with the families you've selected? If they can't, you'll see an element of pain between the eyebrows, reflecting a problem with your definitions. Subdivide the families, stopping as soon as you get to the level where the operating people can relate with confidence to the divisions. In our experience, the finance group, although equally important as the other members of the general manager's staff, tends to be more flexible in handling the definition of families. Generally, whatever marketing, manufacturing, and engineering agree to, finance will accept.

Service Parts

Service parts in some companies are included in the product family they support. In situations where service parts are dominant or the demand for service parts is significantly different from that for the finished product, it is sometimes necessary to create a sub- or separate family. For products that are no longer produced but are still supported, a service parts family is of course required.

In some companies where service parts are not significant and it may be difficult to relate each part to a family, the solution may be to lump all service parts into one family.

Emergency Business

Many companies offer make-to-order products and base their delivery promises on the availability of capacity and materials. Yet, these same companies typically get a steady stream of "emergency" orders that cannot wait for the normal delivery time, but rather must be rushed through on an expedite basis. These situations create much chaos and extra costs. To satisfy the customer who is in trouble, emergency orders are treated as special, given high priority, and rushed through the process ahead of older orders. In fact, they generally disrupt an equal number of regular orders, causing them to miss their delivery promise, because the expedited orders consume scarce capacity and materials.

On the surface, this situation seems like an impossible dilemma. On the one hand, you want to help any customer who's in trouble, and recognize that if you can come through, you'll no doubt be viewed as a good supplier and possibly gain some additional business. On the other hand, you know that it's unfair to delay the other orders that were promised in good faith; you run the risk of aggravating part of your customer base and losing future business. The situation is compounded whenever there is little or no repetitiveness to the individual items. If there were a predictable pattern, you could forecast the individual line item that is apt to become an emergency order, so that it would be available when the customer wanted it. What can be forecasted is the volume of traffic for emergency orders. It's much like absenteeism or scrap. You will not be accurate in predicting who will not show up for work tomorrow, yet there is a pattern as to what percent of the work force will be absent on a day-to-day basis. Likewise, you will not be correct in predicting what item will be scrapped tomorrow, but again good hindsight allows you to predict what volume of material will be rejected on a day-to-day basis.

Companies can analyze their emergency business in a similar manner, treating such orders as a family. Sales and marketing takes the responsibility for forecasting emergency orders as they would forecast any other family, while manufacturing and engineering must respond in terms of allocating adequate resources to service the orders in a "fast track" manner when they arrive. Capacity is thus planned ahead of time. Many companies find that their emergency business comes from certain customers as much as (or even more than) it comes from certain products.

As emergency orders arrive, it should be an order entry or customer-service responsibility to determine which ones deserve special treatment and which ones don't. Furthermore, once the capacity set aside for emergency business in a given period has been consumed, sales must recognize that any more promises must be made in a subsequent time period. The other alternative is for marketing and manufacturing to explore a trade-off where an existing order may be bumped to a later time period in order to insert a newer order earlier.

If the amount of time required to service emergency business is less than the cumulative material lead time, then the forecast must also apply to lower-level materials in order to make them available perhaps at a purchase or semifinished level along with the needed capacity. Obviously, capacity without material is as ineffective as having material without capacity.

Ollie Wight demonstrated this when he consulted with a company that was constantly frustrated by their problems with handling emergency business. Their long-range plan was to wait for the phone to ring. When it did, they slapped on their fire-fighting hats and responded to the crisis by knocking aside regular jobs. Ollie commented, "The phone is going to ring, so why not plan ahead before it does? If you're in the emergency business, why not be the best at it? And being the best means not only responding quickly, but also means not at the expense of existing orders."

This type of emergency planning, which enables you to offer a fast track through your operation, can be a competitive weapon when used properly. A formal system can enhance such flexibility for emergency service if you think through how it should be set up. Unfortunately, many companies think that by going to a formal system such as MRP II they will lose their flexibility, because everything must be done by the numbers. In reality, MRP II will be as flexible as your operation; you need to reflect on what you can do and need to do, and then use MRP II as an effective tool for satisfying those needs.

4. Planning Horizon

The term "planning horizon" refers to how far ahead you need to establish your plans. Everyone recognizes that Sales & Operations Planning is long-term, but the word "long" needs to be quantified. The Sales & Operations plan must extend far enough into the future to ensure the availability of all resources. Thus, whichever resource—material, equipment, people—takes the longest determines the length of the planning horizon.

It's sometimes difficult to analyze which resources will require the longest planning horizon. Additionally, it is not unusual for lead times to change periodically for a particular company. For example, a company requiring highly skilled operators located in an area where the skills are scarce will likely identify that resource as the main constraint to increased output. If, however, there is a downturn among other companies in the immediate area and the availability of skilled people increases, procuring essential raw materials may become the process requiring the longest lead time. There are also times when unusual reactions are appropriate for taking advantage of unusual opportunities. If a rapid response is required to gain additional and profitable business, the company may change its existing policies and subcontract some of the additional work load, even though it may have to pay premium prices to do so. Extraordinary steps may be necessary even if the cost of them exceeds the revenue. One company, for example, decided it was better to pay a "pound of flesh" to procure material from a competitor in order to temporarily protect its position in the marketplace. The alternative of saving money in the short term was judged as being far outweighed by the long-term consequences.

Each company needs to estimate how long it takes to acquire additional resources in a routine manner, and periodically (perhaps every several months or when triggered by specific events) ensure that an adequate planning horizon is in place. Required resources include: skilled workers in the factory, both direct and indirect; skilled office people (for engineering, marketing, etc.); equipment; facilities; tooling; warehouses; new suppliers; subcontractors; and money. Finally, the need to perform what-if or risk analysis pushes out the planning horizon even further.

The what-if analysis applies to each of the aforementioned resources. If a company waits until it must start the execution process in order to provide the resource economically, there will not be sufficient time to study alternatives. Identifying the choices and evaluating the consequences requires considerable time.

Figure 2.4 shows an example of a company in which there is seasonal demand, and its manufacturing organization is attempting to maintain stability. Even though the longest lead time of any resource for this company is six months, the need to look at alternatives or do what-if analysis may extend the planning horizon well into the next year. This is necessary in order to plan production and inventory levels.

Regardless of the fact that longer planning horizons provide greater visibility into potential problems in the future, most people resist making the effort to extend their time line. Not only does it take a lot of work to supply the information that far out, but everyone is aware that such long-range analysis is risky; the crystal ball is inaccurate even in the short term, and quickly clouds up beyond. Still, it's better to go through the effort, expecting some poor guesses, since the process will also yield good guesses as well. Also, to avoid the risk of sometimes being wrong, many companies unfortunately elect simply to be the victim of what happens, instead of influencing the course of events to

Figure 2.4

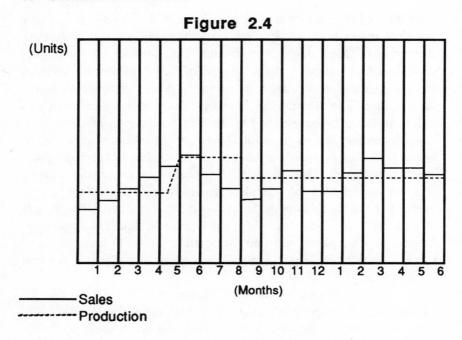

(Units)

(Months)

——————— Sales

- - - - - - - - Production

whatever extent is possible. Most companies operating Sales & Operations Planning will have a horizon of at least a year, with a significant number going out two years, and a few with particularly long lead times going out three to five years.

5. Time Fences

All departments must recognize in their Sales & Operations Planning process that changes in the plan are time-dependent; that is, the closer in the change, the more costly or impossible it may become to make changes in the plan. For every product family, there are "time fences"— guidelines that demarcate when changes are feasible (see Figure 2.5). The fences reflect the realities of each business. As shown in Figure 2.5, the outer fence, *B*, represents the point beyond which changes can

Figure 2.5

Time Fences

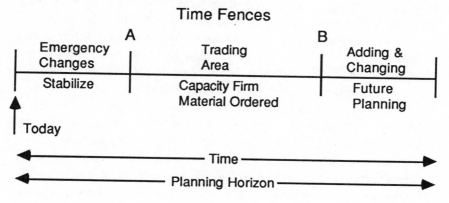

be most easily made. In the middle area between fences *A* and *B*, materials have been ordered, capacity has been established, and changes in rates may consequently be difficult to make. You must, therefore, review rate changes carefully at this point. Changes in priorities, on the other hand, are commonplace, caused by customers changing their demand and forecasts becoming inaccurate. Changes in priorities are easier to implement than changes in output rates. Care is still required to ensure that they can be executed, but generally great flexibility exists as long as the rescheduling is balanced so that as many reschedules out occur as reschedules in. Within the near time fence, even small changes become very expensive as you begin to pay overtime for labor, premium prices for raw materials, and premium shipping charges. Worse still is when you find that you are unable to carry out the change in spite of extraordinary efforts and costs.

For each product family, you therefore need to establish guidelines for determining where changes can be easily made and where change must be evaluated in terms of extra costs. Such time fences can be arrived at by looking at your constraints in terms of capacity and materials.

Many people balk at the idea of time fences, seeing them as straitjackets or barriers. To the contrary, time fences are merely milestones that identify varying degrees of flexibility. This in turn helps you run your business more effectively. You will occasionally have to make changes even though you are within the time fence, but everyone should understand that such changes are exceptions, and if the exceptions become the rule, the cost of doing business and satisfying customer demands will rise sharply. All departments should therefore come to realize that time fences represent the realities of running the company, and gear their planning efforts accordingly. When you need to override the guidelines, it must be done knowing the consequences. Ignoring the impact of changes not only could be costly, but could result in just the opposite from the improved customer service that you seek.

The need to establish time fences also exists at the master schedule level. The thought process is identical, but instead of dealing with changing rates of output, as you are at the Sales & Operations Planning level, at the master schedule level you're dealing with changes in priorities. A great deal of flexibility is possible in the midterm range at the master schedule level if there is a reasonable balance. When there are as many de-expedites as there are expedites, the master schedule is usually attainable, since the changes often are offsetting in terms of their impact on capacity. Although a lot of changes present challenges, whenever a balance between the two directions is achieved, it increases the opportunity to handle changes economically.

Rate changes are less flexible than priority changes. Companies need to analyze how quickly they can increase or decrease output rates so that time fence milestones can be an integral part of the decision-making process. This is not the same as driving an immovable stake into the ground, but rather represents your best estimate of where reactions to altering the plan differ.

Establishing time fences is not an easy task. Marketing, for example, will press hard to have them as close to today as possible, reflecting their desire to service the customers quickly. Manufacturing, engineering, and purchasing, on the other hand, will be advocating time fences as far out as possible, reflecting the fact that it takes time and money to change, and more time means less money. Without guidance from the general manager, an endless debate among the department managers will likely occur. The general manager needs to be actively involved in order to prevent this.

When George Bevis was the senior vice president at the Tennant

Company, he described how he created time fences. Bevis got all of the parties in one room and announced that no one was leaving until they had agreement on where to set the time fences. Each department presented its case and, as expected, marketing and manufacturing were miles apart. After acknowledging that each department had valid reasons, Bevis then agreed with manufacturing's recommendation, with one important condition. "Initially, I'm going to agree with manufacturing," he said. "However, once we demonstrate that by sticking to the time fences we can ship product on time economically, then manufacturing will be expected to work with increasingly less time. We don't know how good we are, but we do know how to figure it out. And no matter how good we become, we'll continue to strive to get even better. My ultimate goal is to be faster and more economical than anyone we compete with."

As Bevis well knew, control and change have equal importance when it comes to understanding time fences. Approving changes that cannot be implemented or can only be implemented at exorbitant costs is foolishness. The Keystone Cops are a good example: much change with little results. The purpose of identifying time fences is to operate with controlled actions. On the other hand, companies must strive for improvements. Shorter lead times have a tremendous payback to all key measurements of the company. They result in faster deliveries, fewer inventories, less obsolescence, less space, less cost, and higher profits.

The time fence issue deserves to be a high priority in any company. Every department should be expected to have an active program aimed at reducing their throughput time. Also, the general manager needs to know each department's goals and to have ongoing measurements reflecting their progress. Don't view time fences as poured in concrete; in fact, they can change as quickly and as easily as the company can. The future of your business is affected by them. If your competition is able to respond faster, it has gained a significant advantage. On the other hand, if you can reduce your reaction times and thereby improve your response capability, *you're* the one who's gained the upper hand.

SALES & OPERATIONS PLANNING AS AN ONGOING PROCESS

A critical aspect of Sales & Operations Planning is that it is an ongoing process, not a one-time event during which production levels are established. For each Sales & Operations Planning cycle, the key players from each department compare actual results to plan, evaluate their performance, and prepare updated plans for the current period.

Sales & Operations Planning meetings should take place at least once per month. A month is usually a sufficient period of time to differentiate a trend from a minor variation or "blip," but it is not so long that corrective action is no longer possible. Still, there are occasions when Sales & Operations Planning meetings must be held more frequently than once a month, depending on the nature of the company and the volatility of the marketplace. Companies that sell seasonal items, for example, might have to meet weekly prior to and during their peak selling periods. Summer comes only once a year; if you manufacture water skis and miscalculate your production needs, you've missed the boat.

Another cause for meeting more frequently or on a special basis is abnormal demand. When demand is suddenly noticeably higher or lower than normal, sales and marketing must decide if the demand represents real change or just a temporary anomaly. Perhaps it represents the beginning of a sustained trend, such as a new product's taking off, or it may represent a temporary situation, as when a competitor experiences difficulty and his customers must look for interim suppliers. In any case, meeting more frequently to analyze the situation will help in understanding the short- and long-term implications of the unusual demand, and it will lead to making an appropriate response. The key point is that decisions about how to handle abnormal situations must be made in a timely fashion at the appropriate management level.

But what about companies that deal in supposedly stable or "mature" markets, or those that manufacture capital equipment with long lead times (airplanes, power-generating equipment, etc.)? Can't they realistically meet less frequently, perhaps quarterly, to discuss sales and operations issues? The answer is a flat no.

Consider the case of a heavy-equipment manufacturer that carried out its Sales & Operations Planning every three months. At one of its quarterly meetings the company's managers were surprised by the bad shape they were in. There had been some small changes in orders for each of the past three months, but by the time three months had passed the changes had become significant, and the required changes in manning levels were severe. Much of this could have been avoided if the process had been performed monthly, because they would have surely noticed that a dangerous trend was emerging and could have taken more appropriate actions to avoid the drastic measures that were ultimately required. The bottom line is that it's better to have a stand-up meeting with little to say than to wait for a meeting in which major surgery must be performed.

A number of companies have paid a hefty price for falling into the trap of believing that long lead times justify long review times. The question is not "How long does it take to make your product?" but "How long does it take to see the need to make changes in your sales and production plans?" Turning a tanker takes longer than turning a small boat. Both skippers, however, need to frequently check their direction to ensure that they are on course and have no obstacles and to initiate appropriate corrections, even though the time required to complete the maneuver is radically different for the two boats.

GETTING SALES & OPERATIONS PLANNING STARTED

Beginning the Sales & Operations Planning process requires a significant effort on the part of the sales, marketing, manufacturing, engineering, and finance departments. As a result, we often hear a number of general excuses for not getting started with the process. While none of them are valid, they can nevertheless become a major impediment to beginning the process. Here are the reasons commonly given for not moving forward:

1. We haven't agreed on the families.
2. We don't have all the data to support the process.
3. We don't have the mechanics to support the process.

4. The general manager doesn't want to look at that much detail.
5. We're a marketing-oriented company, and the general manager expects manufacturing always to say yes.
6. We can't work as a team; I can do it but the others won't.
7. Sales and marketing are in physically different locations from manufacturing—it's hard to get together.
8. What? Another meeting? My calendar is already full!

We encountered a general manager who voiced several of these objections and was as a result extremely leery of beginning the Sales & Operations Planning process. In his opinion, each was a significant barrier and had to be resolved before bringing the group together and conducting formal meetings. We recognized, however, that this would be the same as never getting started. We also recognized that part of his reluctance was fear of the unknown. And even though some of the problems were legitimate, we had to convince him that they could be solved once the group got together.

"Look," we said, "it's like deciding to have kids—there's never a perfect time. If you wait until you're 'ready' to start Sales & Operations Planning, you may never do it. Besides, even if you have everything in order, it's going to take a number of sessions to really refine the process. So get started *now!*"

He agreed but asked, "What do I do now?" We gave him a rough draft of this book so that he could get the process rolling. He also sent a memo to his staff outlining what needed to be done. Before the first meeting, each Sales & Operations Planning team member was required to review a video session of Sales & Operations Planning[1] and read the rough draft of the book in order to be prepared.

At the first meeting (which wasn't really a Sales & Operations Planning meeting—it was a Sales & Operations Brainstorming meeting), everyone discussed how Sales & Operations Planning would impact their specific departments. The issues were dealt with one by one until all participants had a good sense of what the steps entailed and what kinds of benefits the company and each department could expect to reap, along with what was required to support the planning process.

Also, everyone was to return to the next meeting with specific rec-

[1] For more information on video education, see Appendix C.

ommendations for how the process could work most effectively. In addition, the sales and marketing people began compiling data for comparing actual to planned sales and for documenting their assumptions about the marketplace. (As you'll see in the next chapter, there are standard homework tasks that all departments must complete before the monthly Sales & Operations Planning meeting.)

At the second meeting the Sales & Operations Planning group finalized the families, and had sufficient data to step through the process. After the meeting concluded, the general manager expressed a worry to us that the process took too long and involved too much detail. We reassured him by saying, "Sure, we know you'd like to have everything boiled down to one 8½" by 11" sheet of paper. But to really get control of your business, in the beginning we strongly suggest that you accept the fact that there might be a fair amount of detail. Then, as you get better at the planning process, reduce the detail that gets reviewed at the meeting if it does not contribute to the overall planning effort. Eventually you'll reach a comfortable level."

Although the general manager recognized that the process wasn't going to be perfect for a while, he did feel better about the prospects because the group had generated a lot of good questions and had made some decisions. Moreover, he had implemented the "ready, fire, aim" principle of action described in the best-selling book *In Search of Excellence,* by Peters and Waterman. We think this is important, because as we've seen in countless companies, action is the key to generating results.

By the third meeting, the company was carrying out genuine Sales & Operations Planning. Within a few months, everyone began seeing positive results as the production was better synchronized to customer demand. Within six months, the company had developed the process to the point where it was obtaining substantial benefits. These included a better handle on inventory and more realistic production plans.

Does all this sound too simplistic? In a way, the process *is* simple. It does take hard work to make Sales & Operations Planning function efficiently, but more than anything else, it's a matter of *attitude.* The barriers to performing Sales & Operations Planning can be overcome by the simple resolution that you *will* overcome them.

SUMMARY

- A business plan sets forth strategic objectives; Sales & Operations Planning updates the business plan against the realities of the marketplace and the internal workings of the company.
- Before doing Sales & Operations Planning, you must be willing to commit time and people to an ongoing process. A core group must attend *every* Sales & Operations Planning meeting with adequate preparation.
- At the minimum, Sales & Operations Planning meetings must take place on a monthly basis. Seasonal demand and periods of abnormal demand may require more frequent planning sessions.
- Sales and marketing often view families in terms of product lines, while manufacturing typically sees them as a matter of process. Frequently these are the same, but when they do differ, you must have a conversion method.
- The only right time to start Sales & Operations Planning is *today!* No company does a perfect job of Sales & Operations Planning from the start; it gets better over time. Be patient with the aging process and you'll taste the rewards of your efforts.
- The planning horizon must extend out at least one year and possibly longer depending on lead times, market conditions, competition, and other factors that affect supply and demand.
- Establish both material and capacity time fences for changing rates of output for each product family; make sure everyone understands the implications of introducing rate changes. All departments should make an active effort to reduce them, and should be measured on how well they are accomplishing this goal.

Chapter 3
Conducting the Process

. . . a business has no "score" to play except the score it writes as it plays. And whereas neither a first-rate performance of a symphony nor a miserable one will change what the composer wrote, the performance of a business continually creates new and different scores against which its performance is assessed.
(Peter Drucker)

The general manager's job is not an easy one. The company's target is constantly shifting because no market holds still for very long. No forecasting system predicts demand unerringly. On the other hand, capabilities and resources are limited, and no factory, engineering department, or vendor has complete flexibility to change up or down quickly with capacity, or to respond quickly with different mixes. Constraining resources include availability of money, equipment, tooling, skilled labor, etc. Put all these together, and you have a moving target, inflexible weapons, and a limited amount of ammunition. Fortunately, you're competing against companies in a similar situation. It's because of this challenge that you need Sales & Operations Planning.

Sales & Operations Planning creates an opportunity for the general manager to gain control of the business. Since it affects how well the business runs, the savvy general manager will do everything possible in his power to take charge of the Sales & Operations Planning process and ensure that everyone in the company realizes its importance. With Sales & Operations Planning in place, the general manager can determine the direction the company is going in and thus become the captain of his destiny.

The following discussion lays out the key tasks and responsibilities of the general manager in assuring that the Sales & Operations Planning process gets off the ground and produces the maximum possible benefits.

STARTING THE SALES & OPERATIONS PLANNING PROCESS

1. *Lead the educational process*. The general manager must be as knowledgeable about Sales & Operations Planning as everyone else involved in the process. If the general manager doesn't understand the process at the same level as other members of the core Sales & Operations Planning team, he won't be able to effectively orchestrate the process. The best way to understand the fine points of Sales & Operations Planning is to be the teacher who explains how the process works. This is especially important for the general manager of a company that has never performed Sales & Operations Planning, or a company that has met on an ad hoc or informal basis to plan production, but now wishes to institute a formal planning system that will drive the whole business.

In some ways, participation in the educational process may seem like a paradox; successful managers delegate as much as possible. But when they have others learn for them, they give up the opportunity to be effective leaders. Besides, Sales & Operations Planning is the general manager's meeting, and he must be on top of all aspects of the process. By being involved in the design and education process, and by reading all the pertinent literature that has been assigned to department heads and becoming familiar with video courses, he'll take ownership in the process. And ownership is essential for success.

2. *Manage the cultural change*. Sales & Operations Planning affects the entire company culture. Done properly, Sales & Operations Planning guarantees better teamwork by breaking down traditional departmental barriers. By having a frequent update of the needs of the company, everyone becomes aware of what must be done, and what can and can't be done by each department is reviewed. Being expected to do something that they know can't be done is frustrating to the individ-

uals concerned as well as to those members of the team who can do their respective jobs.

The process also affords the opportunity to share in developing alternatives. Personal recognition is one benefit, but equally important is the fact that more creative minds are being brought to bear in determining the best overall path to follow.

All of the above objectives have always been the stated intentions of every manufacturing company. What the Sales & Operations Planning function offers is a means to achieve them. The "doers" thrive in such an environment, while the "talkers" don't. One executive explained, "In our company, if you say you can, and then more than once you don't, that's the end of your career path." This may sound rather harsh and overstated, but the message needs to be loud and clear: don't make commitments you can't back up. Appearing as a "can-do" executive is important, but making promises you know you can't keep can mislead the entire management team and cause serious problems. You may not always be able to come through when you promise something, but if you know you can't do it, don't say you can.

A corollary to the above issues is that the general manager must insist on a culture based on trust and honesty. If you're on a football team and don't believe the people on your left and right, you'll never be able to concentrate on the real competition—the guys across the line. This, of course, is the antithesis of teamwork and leads to a weakened position in the marketplace.

The honesty issue also extends into the area of forecasting and planning. The culture must not reward people who exceed the forecasts merely because they "low-balled" the initial numbers. This charade does not improve the company's competitive posture or financial condition, and in fact may hurt them. In addition, it may evoke deep resentment from other people in the company who will be involved in reacting to real marketplace demands.

3. *Accept the detail.* Most general managers want information reduced to summary form. And that makes sense. How can you run a large company while you're counting the bolts on Model F Widget? Nevertheless, the business runs on detail, and as we mentioned in Chapter 1, the general manager will have to accept a certain level of detail in

order to get the job done. It is best if at first he reviews more detail; then he can gradually cut back to the level where he can still make informed decisions. Similarly, the Sales & Operations Planning team may have to review more families in the beginning than it would like.

Availability of detail is important for backing up the aggregate data when a question arises or an issue needs clarification. The general manager needs to ask for and receive the detail that supports the overview information he's examining. Think of it as peeling back an onion. As a rule, the general manager addresses the top layer, but when it's important to consider more detail, you peel back the onion—the detail has to be there.

4. *Approve procedures*. The general manager should give his seal of approval on a number of procedures used in the Sales & Operations Planning process. These include defining families, determining units of measure, determining the planning horizon, establishing the frequency of review, developing an agenda for the meeting, establishing accountability of performance, presenting proposals, reviewing alternatives, and publishing the results. While some of these items seem mundane, they are nevertheless critical if the planning process is to be successful.

5. *Call the first meeting*. As with almost any activity in life, you can sit around and talk about doing it, analyzing what you need and where it will lead you. But there comes a moment when you simply have to put aside all thoughts and let yourself be driven by your resolve to do it. In the case of Sales & Operations Planning, the general manager must be the catalyst that pushes everyone else into action—he's the only one who has the power to overcome the inertia problem.

Some general managers worry about their ability to lead the initial meetings, fearing that the meetings might degenerate into debate sessions, or that they might be unproductive. Well, that is all likely to happen. At first there will be some scraped and bruised knees, and it may well be that you won't accomplish as much as you would like. If things get tense, call a time-out. Find out what went wrong. Fix it. Then bring people back to finish the job. As turf and other issues get resolved, the meeting participants will function more as a team, and productivity will increase.

While the process of developing Sales & Operations Planning requires a lot of work, starting the process is really simple—call a meeting! The first meeting won't be as productive as future ones, but you'll have broken the ice and will be able to begin assigning tasks for each department. The only way to begin this process, thus, is simply to do it.

6. *Be there.* The general manager's presence is critical while the Sales & Operations Planning process is developing, because it sends a message throughout the rest of the company: Sales & Operations Planning is serious business. By attending classes, video training sessions, and other educational functions, the general manager communicates the sense of urgency that Sales & Operations Planning deserves. If the meeting and education are important enough for the general manager, they will immediately be seen as important enough for other managers, too.

"Being there" also ensures that the monthly meetings are taken seriously. The general manager must be present at all meetings and make it clear that interruptions are to be minimized. While each general manager will have his own style of conducting meetings, the importance of the process must be maintained at all costs.

Beyond the image issue, there's the basic fact that the general manager simply *cannot afford to miss the meeting!* As we said, this is *his* meeting, his opportunity to get control of the business. What stronger incentive does a general manager need for "being there"?

THE ONGOING PROCESS

Once that vital first Sales & Operations Planning meeting has been called, the general manager still has a key role in ensuring that subsequent meetings are productive and that the process becomes more refined with each session. The following suggestions will help in that regard:

1. *Insist on everyone's participation.* There's a danger that after the process starts, people will delegate their responsibility, especially during stable periods. The general manager must not allow this to happen, as it will cause the process to lose its effectiveness. Details may be

missed, and the Sales & Operations Planning group may soon find itself in fire-fighting, rather than fire-prevention, mode. Again, the solution to this problem is to set an example by showing that top management takes Sales & Operations seriously and all other levels of management are expected to do the same.

2. *Make decisions.* The process won't be effective unless the general manager acknowledges that the buck stops at his desk. Often, tough decisions, involving such matters as layoffs and overtime, the timing of a new promotion, and new product introductions, will have to be made to accommodate the revised plan. The general manager must be able to make these decisions to demonstrate his ability to steer the company ship through dangerous shoals.

Moreover, the people in the meeting will take a reading on the general manager's decision-making process. If he's indecisive, he won't be an effective leader. The best approach is to display a take-charge attitude; there is great value in the general manager's willingness to step forward and, when necessary, carry out "damage control" to restore the company when it is jolted by an unexpected event. This approach will be understood and appreciated by all members of the planning team.

Although we have described the process as a democratic one wherein all parties are making contributions, decisions should not be arrived at by counting votes. When we've seen companies operate in this manner, problems are invariably—and unnecessarily—generated. Voting creates winners and losers. And since no one wants to end up on the losing side, lobbying efforts are apt to occur before the meetings, in which individuals line up their votes to go into the meeting with as much clout as possible. This is not the type of homework on which energy should be spent. The general manager needs to hear what each department manager thinks is right and best for the company, regardless of whether it is what the majority thinks.

Additionally, the number of participants attending the meeting may not be balanced per function; some departments may well have more people in attendance than others. Counting votes is hardly fair if this is the case. Finally, not all votes are equal. One general manager summed up this reality by saying, "Before we arrive at a decision, I ask each of my staff members for their input. Almost always, I agree with the majority. When I don't, however, I simply weigh their input. Mine always weighs more!''

A general manager constantly faces many tough decisions. Making the right ones is never easy, but on top of this there are frequent occasions where he must make unpopular ones. That's how it should be. It may be desirable to have harmony, but it's absolutely essential to have consensus, even if it's mandated by the general manager. It's far better for the general manager to be decisive, prepared to make unpopular decisions while still insisting that everyone support them with 100 percent of their effort, than to vacillate whenever there's controversy. A quarterback in a huddle may ask for suggestions, but once he's called the play, it becomes everybody's job to execute it in the best possible manner rather than to second-guess it.

3. *Insist that homework be done.* A Sales & Operations Planning meeting isn't like Hyde Park in London, where anyone can just get up and speak his or her mind. Sales & Operations Planning works only when there is adequate preparation. The need for each department to do its homework before the monthly meeting does not disappear with time. The job, however, gets easier as the departments refine their ability to analyze the past and project into the future. The general manager must be able to quickly recognize when people are walking into the meeting less prepared than they should be, and lay down the law for those who seem to be slipping. There should be *no* surprises, such as a major forecast change that could and should have been reviewed prior to the meeting.

4. *Encourage realistic improvement.* It's fine to have "stretch goals," such as large increases in sales and productivity. But when it comes down to introducing new products, bringing new equipment on line, solving a quality problem, or increasing market share, you'd better be making plans for the company's promises to your customers on the basis of what you *can* do, rather than on what you'd *like* to be doing. The general manager must encourage the representatives of each department to tell him what is really happening in the company and what *can* be done, not what they think he'd like to hear. An excellent way to monitor plans to see if they are realistic is to track whether they are consistently high or low. If they are, then someone needs to attack the cause.

5. *Resolve conflicts.* In any company carrying out Sales & Operations Planning, the general manager serves as an arbitrator. It is inevitable that during periods of unusual changes in demand, sales and marketing, engineering, manufacturing, and finance will have conflicting views about how to respond. Such conflicts are healthy if they are dealt with in the right way, because they lead to a plurality of views. The more views you can generate on a problem, the broader your perspective and the greater the chances for creative and innovative solutions.

The general manager, having studied all the alternatives, is the final arbiter of all major differences in opinion. While he should encourage departments to work out the fine points themselves and learn to compromise, he must be available and willing to hear all sides and make a judgment call. This objective posture is extremely important, because it will encourage active participation by the entire planning team.

6. *Act as a consensus maker.* In the Sales & Operations Planning session, the various departments will be very protective of their interests. Their representatives will also be convinced that the others don't fully understand their unique needs, and that their explanation will not be accepted when they state what they cannot do. The general manager must dispel this notion by accepting what is presented or asking for further information, rather than simply overruling an explanation without clarification.

The need for consensus is paramount. If the Sales & Operations Planning process does not arrive at an approved plan, or if the approved plan is unrealistic and can't be supported, the company may find itself racing back to the informal approach, which includes using hot lists and red tags, expediting, and other inefficient actions. The only way to get out of that style of management and gain control is to do a realistic job of Sales & Operations Planning. And that can come about only through consensus.

7. *Recognize the opportunity for management development among participants.* The Sales & Operations Planning process is not only a better and more effective way to manage the business, it's also an excellent way for the general manager to assist with the growth of his people. There's hardly a better opportunity for the general manager's group to observe each other in action and profit from their observations. The general manager may want to do one-on-one "counseling"

outside the meeting if he's addressing attitude and performance with a specific person. There are times, however, when he must state his expectations during the meeting. His suggestions for improvement can be directed at individuals who may arrive ill-prepared, interrupt frequently, or who question other participants' motives. At other times, he may want to use the meeting as an opportunity to share with the group the types of actions he wants to encourage versus those he wants to be off-limits.

The general manager always needs to recognize that his words and actions are examples for his entire staff. Furthermore, the forum permits him to judge each of his staff in action, evaluating such skills as the ability to analyze situations, communicate clearly choices and alternatives, and interact with the other members of the group.

BARRIERS TO EFFECTIVE MANAGEMENT OF THE SALES & OPERATIONS PLANNING PROCESS

We've identified three major stumbling blocks that general managers encounter in their effort to lead the Sales & Operations Planning process. Read through the following points, and consider whether they might apply to you. If so, how might you change your outlook on the planning process?

1. *Fear of detail.* As we previously mentioned, good managers prefer summary data to reams of raw detail. But as we also explained, some degree of backup detail is necessary to make the planning process work. Don't be afraid of being inundated; after all, you have the control to force others to change the detail/summary ratio on any report that's handed to you.

2. *Bias toward one department.* If the general manager has worked his way up through one of the departments, he must take extra care to ensure that he has a balanced view. Any biases will instantly breed resentment and a divisive atmosphere, which are guarantees that the planning process will be short-circuited. The best meeting leaders are tough but fair, and work hard to monitor and check their own personal leanings.

3. *Lack of understanding of the process.* Some general managers never really understand the benefits of Sales & Operations Planning because they haven't taken the time or initiative to learn about the individual mechanics, let alone the process as a whole. As a result, they just aren't motivated to provide the level of leadership necessary to make Sales & Operations work. Sadly, these people also lose their best shot at gaining control over the business, and never learn to manage change effectively.

SUMMARY

- As with any process that involves top management, the general manager must evoke the respect of his staff by demonstrating a keen understanding of its mechanics and the ability to make tough decisions.
- Sales & Operations Planning works when you have a coach, not a boss; a leader, not a dictator; a resolver, not a "ducker."
- An effective general manager has four-way vision: he looks backward to evaluate performance, forward to see what lies ahead, upward to see the strategic plan, and downward to examine the detail plan.

Chapter 4

Establishing Responsibilities

You know why composers live so long?
Because we perspire so much.
(Sir John Barbirolli, British conductor)

THE REALITIES OF THE PLANNING CYCLE

Would that we could make time stand still at the end of each month! We'd be able to complete all the steps that are required for the Sales & Operations Planning process, then start the clock and slip into the next month. Alas, the real world isn't so tidy—it's impossible to stop time or collapse the planning process down to zero. The more time you take going through the planning steps, the less time you'll have to put a revised plan into action and adjust for changes in demand (see Figure 4.1).

Conversely, the more quickly the planning process can be concluded, the less overlap you'll have between periods, and the more accurately and timely you'll be able to formulate an appropriate response to current demand. This increase in speed and efficiency is accomplished through better orchestration of the whole Sales & Operations Planning process, which enables each department to respond to current data as quickly as possible and develop a set of recommendations for dealing with changes in demand.

To conclude the process the following milestones must be achieved:

Figure 4.1
Sales & Operations Planning Process

Last Month	This Month	Next Month
	End-of-Month Processing	
	Update Sales Plans	
	Update Engineering Plans	
	Update Manufacturing Plans	
	Update Financial Plans	
	Monthly Meeting	
	Execute Plans	

1. *Complete end-of-month processing.* Accounting and data processing close out the previous month to provide each department with the appropriate data, performance information, and status information.
2. *Evaluate performance to plan.* Each department must evaluate its performance to plan in the past period. Any significant differences must be investigated, and the causes identified.
3. *Update plans.* Each department updates its plans, not only on the basis of past performance, but based on the input as it flows from other departments.
4. *Gain company-wide approval.* All departments meet to discuss the impact of changes and agree to a company plan.

Based on our experience, the length of time required to complete the process depends on the company in question. Companies that have fine-tuned the process find that they can conduct their Sales & Operations Planning meetings by the tenth working day of the month. While it is desirable to hold the meeting as soon as possible, it is equally critical to ensure that the proper homework is done and that the timing is consistent from month to month. In this chapter, we'll describe what each departmental response entails, and explain how the necessary data can be assembled most efficiently to keep the Sales & Operations Planning process productive.

SALES & MARKETING'S PRE–SALES & OPERATIONS PLANNING RESPONSIBILITIES

Sales and marketing have the responsibility to do whatever is necessary to develop a statement of demand at both the detail and aggregate product levels. In order to help determine the best statement of demand, sales and marketing must determine which approaches and tools are most appropriate for the various product families. The first step is to determine whether to forecast and plan in aggregate and then break the forecast down to the detail, or to forecast the detail and aggregate it into a family. Either approach can be appropriate, although the sequence in which they are done can change over time. Most important is that the aggregate and detail projections be linked together in a closed-loop approach.

Sales and marketing must understand the level and patterns of demand for the company's products before they can predict customer demand. Most companies have a mix of product types, each with its own demand pattern. For "stable" items, you will be relatively safe in using the past to predict the future. But what about "life-cycle" products that have passed the growth phase and are on the decline? If you look at past performance of life-cycle products alone, you might deceive yourself into thinking that the next period will be a continued increase, when in fact the product may have peaked and your company should plan on diminished sales from that product line.

Seasonal products present unique problems because they usually require a buildup of inventory. The point is, you must examine your product mix and organize products into appropriate families so you can analyze the past and project the future. (For a complete discussion of understanding and managing demand, see Oliver Wight Limited Publications, *The Marketing Edge*, by George E. Palmatier and Joseph S. Shull.)

Sales & Operations Planning in a Distributed Product Environment

In order to get close to their customers and minimize transportation costs, many companies use distribution centers. In those cases, the aggregate of all of the forecasts and inventory information related to the centers must be included in the Sales & Operations Planning process. Distribution Resource Planning (DRP) is the recommended approach for managing this function. (For a complete discussion of this topic, see *Distribution Resource Planning*, Second Edition, by André Martin.)

Planned versus Actual: Performance Evaluation

The first step in reviewing past performance is to look at your own sales and marketing efforts and results. Where actual did not equal plan, was it something you did or didn't do that caused the deviation? Was a new promotion exceptionally good or bad? Was the timing of the advertising and promotion on target? In short, did sales and marketing do a better or worse job for the last period than in the previous periods? This determination requires unbiased judgment on the part of management to evaluate the sales effort.

Evaluating Change

Significant deviations from the sales plan need to be investigated. Deviations can be separated into three categories:

1. *Volume deviations.* Customers order more or less product for the entire product family.
2. *Mix deviations.* Deviations occur for items within a family where there is no volume change.
3. *Timing deviations.* The volume and mix may remain the same over time, but there is a change in the timing of orders.

Unfortunately, it's not always so simple to identify which of the three types of deviations is occurring, because there could be a combination of events. The most critical of the three is volume change, because it can carry with it capacity ramifications for engineering, manufacturing, and purchasing. Making the decision to change aggregate rates at the right time is the essence of a well-tuned Sales & Operations Planning process and a well-managed manufacturing company.

Deviations in product mix are also important and can be as troublesome as volume changes, because they affect your ability to satisfy your customers and can result in major scheduling changes. Moreover, if the change in mix is significant, the result can be excess inventory and back orders in a make-to-stock environment and an extending backlog in a make-to-order environment. Sometimes, in order to satisfy customer service problems, production needs to be increased temporarily until the inventory and backlog get back within tolerance of target levels.

Finally, there are timing deviations, which over the long term may not represent real changes at all. Timing changes are sometimes the most difficult to identify because they are easily mistaken as volume changes. If you respond to timing changes by increasing or decreasing capacity, the results might be harmful. For example, let's say that you sell through distributors, and several coincidentally decide to order during the last week of a month. This could give the appearance of a surge in demand, while in fact it doesn't represent a real increase. The worst possible response in this situation would be to change the plans for the affected products, because the company would slip out of sync with the marketplace, causing a misuse of capacity and an excess of inventory.

Timing changes are likely to come about when your company's demand is dominated by a single customer. Let's say your forecast is for 120 widgets per year, which breaks down to 10 per month. Your main customer orders 25 widgets. That may look like a significant deviation, but in fact it may simply be an order for a three- or four-month period. Your company would be overreacting if it interpreted this order as additional demand or an upward trend. This problem can be dealt with by either "linking" with the customer or separately forecasting and tracking that customer's demand.

Figure 4.2

	1	2	3	4	5	6
Old Forecast	100	100	100	100	100	100
Actual Orders	118					
New Forecast		100	100	100	100	100
No Change Forecast		82	100	100	100	100

Companies often fail to understand how to communicate "no change." Many people think that if the new forecast is the same as the old forecast, that means "no change." The example shown in Figure 4.2 helps illustrate the concept. Last month, the forecast was oversold by 18 units. A real no-change forecast would have to subtract the excess demand from the future forecast, as shown in the no-change forecast row. But if you look at Figure 4.2, you will see that this was not done. Instead, this communication has just told manufacturing that the forecast of demand for the six-month period was increased from 600 to 618. Such simple misunderstandings have caused a lot miscommunications and second-guessing on the part of manufacturing organizations.

It is thus vitally important to determine whether deviations are real, whether they represent timing change, whether they are abnormal demand requiring temporary resources, or whether they constitute a bona fide volume change that may be the first indication of an increase or decrease in business.

Abnormal demands that are one-time and nonrecurring may stem from a number of sources. For example, a surge in demand might be caused by a competitor who's having temporary problems. If that's the case, enjoy your "opportunity orders," but don't necessarily count on sustained demand. Your competitor's customers may resume their bonds of loyalty once the problems are overcome, so it may be unwise to change your capacity or give the customers a priority in response to what is in essence a short-lived phenomenon.

The next step is to determine which tools and techniques will best serve the product environment. Possibilities include statistical analysis to forecast trends and/or seasonality, direct customer linking, which will eliminate the need to forecast, or tracking economic indicators that inform you of the changes you are about to encounter.

Whatever approach you use requires an effective measurement process to see if you're obtaining the desired results. Measurement is sometimes a controversial issue, since people are often overly sensitive and defensive about being measured. The proper use of measurement, in which forecasts are measured against actuals, is to point out areas that need improvement. This also informs management of those areas in which better forecasting performance is impossible, so that it will be prepared to cope with forecast deviation.

Analyzing and Documenting Your Assumptions

Behind any plan or forecast, there are underlying assumptions concerning factors that will affect your business. It's important to identify or state these assumptions, because they may explain why the numbers have changed and whether the change is likely to be short- or long-term. Many companies don't do this, and therefore don't understand the foundations on which their plans are really built. This makes it impossible to get a complete grasp on what went right or wrong with your original thinking. Perhaps you based your original sales plan on a certain exchange rate for the dollar. All of a sudden, your currency increased in value and your product became too expensive for foreign markets, so you found yourself with a drop in sales. The problem isn't your forecasting, but rather factors beyond your control.

A similar scenario could also be painted for the price of oil, interest rates, inflation, etc. The point is, if you did a solid job in documenting your original assumptions, you will have less difficulty tracing the logic supporting your plans and making adjustments to it, and you'll know why actual performance deviated from the plan.

The intent of documenting assumptions is not to put the blame on anyone if the prediction didn't turn out to be accurate, but rather to be able to identify the cause of the deviation. This understanding can contribute to correcting the cause and thereby minimize its recurrence.

Even if it's not preventable, the explanation will add to everyone's grasp of why the unexpected happened. This knowledge will strengthen the planning process; the absence weakens it. Worse still, you may misjudge a situation as resulting from poor performance, when in fact it was beyond anyone's control.

The identification of assumptions also enables everyone to recognize what they're "signing up" for. Better to proceed recognizing these factors than to approve a plan when unaware of them, only to encounter problems at a later date.

The following list includes factors you should consider. Some are hard or impossible to control, yet they must be closely monitored because they can have a significant impact on your business. Regardless, they should be covered in your list of documented assumptions.

Types of Assumptions to Be Considered

General Economy: What are the factors that affect business in general?
• political events (general elections, etc.)
• inflation
• interest rates
• exchange rates
• commodity prices

Market Outlook for Your Types of Products: What are the factors that affect your product lines?
• market size
• new technologies
• consumer preference
• regulation

Market Share: Which factors determine what percent of the market you're able to capture?

• competition (who's competing, their pricing and advertising strategies)
• your strategy (pricing, marketing campaigns, sales and distribution coverage, new products)

In many situations, it's the assumptions behind the forecast that are wrong. Yet how many times have we blamed the forecast and the forecasters when actual demand is different from the forecast? An understanding of why and how to document assumptions can lead to a better and healthier attitude toward forecasting, which will result in better forecasting performance.

Determining Where You Are Vulnerable

Now that you have documented the assumptions about factors that can affect your business and are reviewing and updating them, it's time to take the next step and determine where your business plans are most vulnerable. What are the factors concerning your business that are the most critical? What can you do to ensure that you can control those factors when possible? If the factors are not really under your control, then what kinds of contingency planning should you perform to help guarantee good business performance? To answer these questions the following areas of vulnerability should be explored:

- internal issues (Will you have sufficient resources in terms of labor, material, and tooling to support the plan?)

- new-product timing (Will the new products be released when anticipated?)

- external issues (Which of the factors beyond your control seems the most threatening: interest rates, exchange rates, competitive pricing, weather, government decisions, elections, etc.?)

The above list describes the downside or vulnerabilities that you might face. But what about the upside? What if there is an opportunity to do more business? Without planning for such situations, you're likely to let them slip by unexploited. This might happen with a seasonal business that exceeds expected demand. Without the raw materials or ability to flex capacity or subcontract work, the opportunity will be lost. And nothing is more frustrating to a sales and marketing department than to see opportunities lost because of lack of resources. Such problems can be avoided with contingency planning.

To illustrate the value of contingency planning, let's consider how four companies employ it to be more competitive in their respective marketplaces.

We saw one company that had a major concern about its ability to react to increased demand on a seasonal product. The company could flex capacity, but couldn't as easily adjust material, so its management began planning for a possible increase by strategically stocking raw material. The first time the company did this it was rewarded with an impressive increase in business for a very small risk investment.

The second company is a high-volume producer of low-cost, short-lead-time consumer goods. Promotions are run regularly, and it is important to respond quickly to orders for advertised items. However, the forecasts of promotions are unreliable. To address this problem, the company schedules all production to be completed within a five-day workweek. Saturdays are used to make more product if actual sales exceed marketing's forecast. The master scheduler is authorized to plan sufficient materials to support this contingency plan. Purchasing has worked closely with its suppliers to minimize the special labeling for promotional material in order to avoid liabilities for the extra components in the event they are not needed.

The third company manufactures a product with the opposite characteristics: high cost, long lead times, very low volume, and all units made to order. Most customers place orders based on standard delivery times. There are, however, "orders of opportunity" where the sales depend on much shorter delivery times. These orders can be very profitable. At the Sales & Operations Planning meeting, management reviews the desirability of scheduling unsold units. Marketing becomes accountable for these units, acting as the customer. It is marketing's responsibility to specify the options and to determine what customer eventually will become the "real" owner.

The fourth company established its contingency plan to coincide with its time fences. Beyond the cumulative material lead time, marketing can request up to 50 percent additional product to protect against forecast errors. At the cumulative material lead time, marketing must decide how much of this they will likely need. Once decided, they are restricted to changes of no greater than 20 percent. At the close-end time fence, marketing makes its final review, and thereafter they are not expected to make any further changes.

Contingency planning provides advanced information to the managers who are involved in implementing it. If the alternative plan becomes the approved plan, they must be prepared to support it. In turn, these same managers can spell out the consequences of what the approved plan means. This permits the general manager to weigh the advantages against the disadvantages and determine the most competitive course of action for his company.

Sales Planning in a Constrained-Capacity Environment

Sometimes sales and marketing find themselves in an environment where the company is operating at full factory output. A capital-intensive chemical operation that is running three shifts seven days a week would be a good example of this situation. In such instances, sales and marketing take on a different role: their job is to determine how to allocate capacity among families and to make sure that the total capacity is sold. The product mix can also be very important in this environment in order to maximize profits while balancing capacity.

What-if Analysis of Demand Alternatives

In the current changing environment of global competition, one of the most essential and powerful tools is "what-if" analysis. Managers have long wished for the ability to simulate the effects of different planning alternatives and options. This is possible with the current generation of computers and software now on the market, and each company should determine what kind of what-if simulation capabilities it needs.

What-if analysis should take place at the Sales & Operations Planning level. The analysis of demand alternatives and various response options can yield a better understanding of both the problems and opportunities facing your company. Much of the analysis can be handled with rough-cut capacity planning, which will be discussed later in this chapter.

SALES PLANNING FOR THE FUTURE

Once sales and marketing have analyzed past performance and assessed whether or not real change has occurred or will occur, their next
step is to evaluate any additional input from the marketplace and use
that information to update the sales plan. Even if no future projections
are changed, the updating will still be an important reaffirmation of
the plan. Without knowledge of why certain events took place in the
past, sales and marketing can't confidently use that information to plan
for the future. The following scenarios illustrate how sales planning
contributes to the process of developing strategic actions.

(Note: In the scenarios we use the terms "make-to-stock" and "make-
to-order." Make-to-stock refers to off-the-shelf products, whereas make-
to-order refers to a product that is either made or finished after receipt of
the customer order. In addition, later in this book you will encounter the
term "engineer-to-order," which refers to a custom-designed product.)

EVENT:	Sales were down.
ANALYSIS:	Unable to hire salespeople. The forecast depended on having 234 salespeople in the field. Our head count for the last period was 216. The shortfall was *real*. We lost a business opportunity because of lack of coverage.
ACTION:	Hired 18 people. We've corrected the problem; therefore we're not reducing the future sales plan numbers.
MESSAGE TO MANUFACTURING:	This has cost us business in previous period. Shipments for the future, however, should be back on track.

EVENT:	Sold above forecast.
ANALYSIS:	Total market was flat, but we improved our business by 33 percent by increasing our market share

from 18 to 24 percent through aggressive pricing. While we expected only a 15 percent increase, we believe there is an even greater opportunity for increased market share—26 to 28 percent—based on improving quality and on-time delivery. This has been validated by a recent customer survey. New product plans are on schedule.

Note: Our documented assumption is that we can improve our quality and on-time delivery performance. Our forecast is that as a result, we will continue to increase market share.

ACTION: Increase the sales plan by 20 percent.

MESSAGE TO

MANUFACTURING: Gear up the factory to correspond to the sales forecast. Advise sales and marketing as soon as the new production plan has been laid out taking into account the impact of manufacturing's ability to respond to the demand. We've got to be very careful to make a real increase in manufacturing, because we are also trying to improve on-time delivery. If we don't plan well, our strategy could backfire.

EVENT: Demand was off by 5 percent

ANALYSIS: The marketplace is softening. Interest rates are up, housing starts down. But we maintained our market share. We expect business to be off unless we can do something to increase market share, but increasing market share is not feasible at this time.

ACTION: Predict that business will be down 5 percent for the rest of the year.

MESSAGE TO

MANUFACTURING: Ramp down slightly. Cut down on overtime and let normal attrition reduce our work force.

EVENT: We hit the shipping plan and bookings are up.

ANALYSIS: The marketplace seems steady; however, the delivery lead time for our make-to-order product is lengthening. Our analysis shows that it is not just customers giving us more time to deliver the product, it's *real* demand. This appears to be an opportunity to increase business for make-to-order, but only if we can avoid long delays. Our current mix is 60 percent make-to-stock, 40 percent make-to-order. We're confident of a 25 percent increase in make-to-order segment, and a 15 percent increase in total business.

ACTION: Project an increase to this level over six-month period.

MESSAGE TO
MANUFACTURING: Handle volume and help prevent backlog from extending too far (beyond a competitive position).

In each scenario, review of the past helped the company to effectively plan for the future, and helped management to make longer-term tactical decisions and, in some cases, even modify strategic decisions. Not all monthly reviews will lead to brilliant insights and major changes. In some cases, there may be no change at all, or the changes may be very minor. Regardless, marketing must communicate the plan for the future to the rest of the company. In fact, the output of sales and marketing's pre–Sales & Operations Planning meeting is the input to the other departments' planning process.

Now that we've covered the basics for sales and marketing, let's see what happens in the pre–Sales & Operations Planning sessions elsewhere in the company.

MANUFACTURING'S PRE–SALES & OPERATIONS MEETING PREPARATIONS

One of manufacturing's normal objectives is to maintain as stable an employment base as practical. This necessitates developing a manufacturing strategy that provides flexibility by using overtime, subcontracting, inventory buildup and depletion, and backlog changes, so that the company isn't in a constant hire-and-fire mode. A good model for this approach is Bently Nevada, a manufacturer of electronic controls. Bently Nevada's goal is to flex output by 25 percent without changing the permanent work force. This is accomplished by the use of temporaries, cottage businesses, and overtime.

Decisions such as those that were implemented at Bently Nevada are made at a top management level and come down to this simple question: "Do we want to have a production schedule cast in stone and expect marketing to bend, or are we looking for manufacturing to maintain flexibility to cope with a volatile marketplace?" Whatever balance your company chooses, it must have a firm understanding of the impact that changes in demand will have on temporary and permanent resources. This is what manufacturing's pre–Sales & Operations Planning process is all about.

During its pre–Sales & Operations Planning process, manufacturing must not only ask, "Can we respond as quickly as sales and marketing would like?" but equally important, they must ask, "Does it make sense to respond this quickly?" From marketing's standpoint it might make sense. But the cost of responding might be prohibitive, and alternatives from manufacturing must be presented so that the general manager can ultimately make a decision.

The Impact of Change

When sales and marketing present their updated sales plan, manufacturing will be faced with one of four situations:

1. *The aggregate plan is unchanged.* This, of course, is the easiest situation and calls for business as usual (unless there is significant mix change or unforeseen support problems).

2. *The plan is downsized.* If sales and marketing see only a short-term dip in demand, manufacturing will likely recommend holding the line and building inventory or cutting backlog rather than laying off people. Perhaps manufacturing will suggest alternative work schedules during the period of reduced demand. To retain a skilled and loyal

work force, some companies put people on reduced work schedules rather than on unemployment when sales are slow. The result is a dedicated work force that is always available when the market resumes.

Another difficulty with downsizing the plan is the effect on suppliers. Unless you communicate to your suppliers why production is being reduced and how long and severe the reduction will be, you may cause future trouble for them. By not communicating with suppliers, a company can set into motion a series of out-of-control actions that will affect the suppliers.

Picture a supplier who guesses that the downturn is only temporary, and there is no need for him to reduce his output rate. As he continues to produce at his existing rate, he is building up more inventory than expected. Eventually, his inventory will reach a level that is unacceptable. At this point, he must take some dramatic steps to adjust production down and work off the buildup. During the same period of time, the buying company, having gone through its downturn, changes its rate of production up and starts ordering materials at the higher rate. The supplier is at the lower rate and, having recently been burned, will likely be reluctant to act quickly.

All of the above problems occurred because of the lack of communication. Good communications would have given the suppliers the choice of how they wished to respond. But regardless of the different actions that each may take, everyone needs to be aware of when the lull will end and higher purchases will start again.

3. *The plan is increased.* This is the biggest challenge, because it poses the most constraints. What does manufacturing have to do to gear up, and how quickly can it be accomplished? How quickly manufacturing can ramp up depends at least on capacity and availability of material. In answering these questions, manufacturing must look at material and capacity simultaneously. Unless you have or can get all the required materials from suppliers, extra capacity is irrelevant. It will be necessary to talk with vendors and discuss ways to get material sooner, or to explore alternative vendors who can meet your timing needs. Likewise, if purchasing can get the material but manufacturing doesn't have the additional capacity, the material is irrelevant. Other potential problems with increased volume could lie in obtaining the required tooling or space required to support the plan. More informa-

tion about material and capacity planning is provided below.

4. *The aggregate plan is unchanged, although there is a mix change.* This type of change can be quite troublesome. You don't want to authorize more production in aggregate, but in order to get the required inventory to provide adequate customer service, there might have to be a short-term increase in production even though you'll have to cut back in the future to compensate. The alternative is to suffer through a short-term decrease in customer service.

Material Planning

The materials questions are assessed through a material-availability and lead-time analysis. This entails looking at the materials that are on hand and on order and then checking cumulative lead times to see how long it would take to acquire more. Cumulative lead times include purchasing and manufacturing times and, if required, engineering time. Rather than simply accept this answer, however, many companies aggressively work at reducing the longest material lead-time path. They do so by periodically identifying what items are in the path, and then analyzing what steps can be taken to cut time out of it. Some of the reductions have been very dramatic in companies pursuing Just-in-Time.

The purchasing lead time is always a significant contributor. Discussions therefore need to be held with the suppliers who provide the items in question and should result in an understanding of your needs, an understanding of their needs, and agreement on what steps can be taken jointly to help each other.

Good communication and good data become good ammunition. Companies that have Manufacturing Resource Planning should be prepared to share planned orders with suppliers all the way out through their horizon. This information enables suppliers to do a better job of material and capacity planning within their companies, and thereby reduce their firmed-up lead times to be more responsive.

The traditional relationship between customer and supplier is one of opposing forces. The customers push hard for the supplier to offer short lead times, while the supplier pushes back for longer ones. A more enlightened way to work together is to discuss how far into the future the supplier would like to see your requirements, and then at what point the requirements need to be firmed up. "Vendor scheduling" is the term used to describe what companies do with the information from Manufacturing Resource Planning to accomplish this interchange. The specifics are described in *High Performance Purchasing,* by John Schorr and Tom Wallace. As soon as some paths are shortened through better vendor communications, others obviously become the gating ones. The process, therefore, must become an ongoing effort.

Another way to reduce the impact on the cumulative material lead time is through "overplanning." This means that the general manager and his staff have decided to authorize a higher output rate than is necessary for what they expect actually to need. The purpose of this approach is to predict what resources will be required in the event that they really are needed. When overplanning occurs beyond the cumulative material lead time, it simply shows the potential impact on capacity and finances. If the overplanning occurs inside a cumulative material lead time, then the long-lead-time purchase parts will be ordered and the manufacturing process started. It is therefore important that overplanning be managed carefully. The general manager and his staff should know the consequences of overplanning, particularly in terms of capacity utilization and inventory investment.

To implement an overplanning decision, the master scheduler is authorized to schedule more products in the designated time periods specified by management in its strategy. As these extra time-phased quantities are exploded, they feed both the material planning process and the capacity planning process.

Where the marketplace is volatile and your goal is to respond quickly to any upturns, overplanning is an attractive strategy. Additionally, if the long-lead-time materials that are being procured and manufactured are popular items, overplanning can be a low-risk proposition. Unfortunately, sometimes many of the items in the long-lead-time paths tend to be low-volume, high-cost items, which when procured and/or made become an expensive cost if the marketplace turns in a different direction.

By reducing the cumulative material lead time and/or by overplanning, companies gain flexibility in responding to upturns in business. This flexibility is obviously a competitive weapon, but like all alternatives, it comes with trade-offs in the opposite direction. The combination of importance and risk means that the decision should be elevated to the Sales & Operations Planning process. It is here that the general manager and his staff need to understand the pros and cons, and then make the decision that best serves the company's objectives.

Capacity Planning

In order to answer the question "How soon can manufacturing gear up and what is the cost?" it is necessary to project capacity requirements. This can be done by (1) detailed capacity planning or (2) "rough-cut" capacity planning. While some companies can do a detailed capacity plan whenever needed, because of their manufacturing environment or a limited number of work centers, most companies will have to use rough-cut capacity planning because of time or complexity constraints. Full detailed capacity planning is normally too cumbersome and lengthy a process. The following section explains the basic rough-cut approach.

Rough-cut capacity planning is the process of determining the impact on key resources required to support the proposed production plan. Rough-cut capacity planning is not solely for manufacturing; engineering and distribution can just as easily use the following principles to determine necessary capacity.

Preparing for rough cut involves the following three steps:

1. Picking Critical Resources
Step one entails determining which resources are most critical in terms of changing the production process. The idea of rough-cut is to be able to look at a small number of key resources instead of examining every resource used in the manufacturing process of a particular product family. Here are some general guidelines for identifying these resources:

A. *It's a bottleneck.* Potentially, any resource (people or equipment) can become a bottleneck and, therefore, a critical resource if it is currently being used to capacity. Perhaps your manufacturing process requires a certain piece of equipment, but you only have one operator who can use it. Even though that equipment could be used three shifts a day, the lack of skilled people who can operate it makes the people the critical resource.

B. *It's difficult to off-load.* Any resource that cramps your flexibility can be considered critical. By this we mean a process that can't be subcontracted or done with alternate equipment. This could include some critical skills.

C. *It requires a long lead time to make a change.* If you can change a resource in a short lead time, you have no problem. You may add a shift, hire more people, subcontract, etc. But if the lead time to change the capacity level (i.e., hire and train people or order more equipment) is many months, and you need a quick increase, you're in trouble.

- *A combination of the above.* Examples of a combination of items could include:
- *A final assembly line that requires skilled people and special equipment.* Subcontracting might be absolutely out of the question. Additional lines would require large capital outlays and you might actually be operating all shifts.
- *A work cell or flexible work center that might be used for short-lead-time production.* Even though it is flexible in terms of what products can be produced, there is a limit to how much total capacity the cell can handle.

In many manufacturing environments the resources that are critical may change from time to time, and must be periodically reviewed. For example, grinding machines may be critical this month, but a mix change could make stamping machines the critical resource next month. In such situations, you must be careful about deciding which resources are likely to be bottlenecks. The bottom line is that rough-cut is *not* a substitute for detailed capacity planning, which takes into account the impact problems of product mix on all work centers. Additionally, rough-cut is generally reliable, as long as you recognize it won't yield four-digit precision or complete coverage of other work centers.

2. Determining the Capacity Impact on Critical Resources

The next step in developing a resource profile is to calculate the impact that the products you produce have on the critical resources you have selected. This is the number of standard hours that will be required for each item to be produced. If the routings and standard hours are known, you might determine the resources required at the family or the master schedule level, as explained below.

A. Family Level
Two approaches are possible at the family level:
- *A single item in the family can be used to represent the entire product group.* This makes sense when all the products in the family are very similar in their use of the critical resource.
- *Calculate an average.* In some cases, no one item in the family is specifically representative, because the capacity content varies. The most effective profile in this situation may be an average that represents the group as a whole, rather than any one particular item. In order to catch changing mixes, you should periodically update this weighted average.

B. Master Schedule Level
It's sometimes difficult to develop a reliable resource profile at the family level, because capacity and volume can vary significantly by individual item. In that case, it's necessary to develop a profile for each item in the family that is master scheduled. This approach reflects which items and what quantity are in the master production schedule.

If your company does not have routings, then the following techniques can be used:

- *An estimate from a knowledgeable person.* If hard data aren't available, at the very least ask a foreman or other experienced person intimately familiar with the production process how much time will be required to make the product. While you may think that such estimates are not precise enough, they will at least give you a general sense of the production requirements, which is sufficient for this process. Remember, absolute precision isn't the goal; rather, the idea is to predict the approximate size and timing of new rates of production.

- *Average hours on like products.* You may not know exactly how many hours are required on model 200, a new product, but you know it's similar to model 100. So you can use model 100 as a reference for the new product.

3. Determining Timing/Offset Lead Time.
Typically, you work from a plan that shows when the product will be produced or finished to support the customer or the forecast. For some

Figure 4.3
Resource Profile

KEY WORK CENTER	HRS PER PC	LEAD-TIME OFFSET (DAYS)
Finishing	3.2	0
Subassembly	2.5	5
Welding	1.6	12
Milling	5.2	18

work centers, the capacity may be needed at one of the beginning operations, and a lead-time offset will be necessary to predict when the capacity is needed.

Performing Rough-Cut Planning

The capacity profile consists of selected critical resources, the standard hours required, and timing. To use them for rough-cut capacity planning, take each of the families in the production plan, multiply them by the profile, and summarize the results by each resource. This will yield the time-phase required number of hours (i.e., required capacity) to support the plan. The result in turn needs to be compared against the demonstrated capacity. This is defined as the standard hours of output that the company has actually been achieving. The objective, of course, is to review significant differences between required capacity and demonstrated capacity. Implementing a revised plan depends on getting the calculation of required capacity and measurement of demonstrated capacity to agree.

Based on this rough-cut information, manufacturing management should step forward and make a recommendation about what can be done. During a period of tough change, when manufacturing believes that the ultimate decision will require more dialogue with sales and marketing and the general manager, it should also present alternatives that can be discussed during the Sales & Operations Planning meeting.

The following scenarios show how a combination of material lead-time analysis and rough-cut capacity planning can be used to aid the decision process in manufacturing:

SITUATION:	Sales and marketing call for a 10 percent increase in our make-to-stock product line.
ANALYSIS:	Finished goods inventory is low—many items are out of stock. Normal material lead-time constraint—ninety days. Checked with vendor to see if can get material in shorter lead time. Answer: forty-five days. Ran rough-cut capacity plan. Permanent increase in machine shop capacity not feasible until sixty days. People constraint, not equipment constraint.
ACTION RECOMMENDATION:	Production will be increased with overtime in forty-five days, permanently in sixty days. There will be continued customer service problems. Look for short-term relief by cutting safety stock quantities to free up capacity. Sales must be made aware of the problem.

SITUATION:	Sales and marketing are projecting an 8 percent gradual increase in a major make-to-order family over the next four months.
ANALYSIS:	Customer delivery objective is four weeks. Current backlog is five weeks. There will be a customer service problem if lead time extends beyond six weeks.
	Manufacturing's objective is to respond to increase in demand, to keep backlog below six weeks and get back to a four-week competitive lead time. Would be desirable to ramp up production faster than sales rate to drive delivery lead-time down.
ACTION RECOMMENDATION:	Material availability checked—available in thirty days. Recommend overtime in month-two time

frame, to be replaced by regular-time 8 percent increase in capacity in third month.

SITUATION:	Forecast down 10 percent. Don't see a pickup in the short term.
ANALYSIS:	Company policy is to avoid layoffs if at all possible.
ACTION RECOMMENDATION:	Eliminate any temporary help. Cut overtime to the absolute minimum. If any attrition, will not replace. Encourage personal and vacation time where appropriate. Build inventory in the short term. Monitor the situation very closely; if it continues, may cut work force back to four-day week. Form a task force to bring in outside contract work.

ENGINEERING'S PRE–SALES & OPERATIONS PLANNING RESPONSIBILITIES

Although engineering is not usually involved as much as sales and marketing or manufacturing in the Sales & Operations Planning process, there are some areas, such as new products, where it plays a vital role. In engineer-to-order companies, engineering's role is equal to that of manufacturing, so engineering must look at the proposed sales plans and respond *first;* manufacturing, of course, follows engineering and is therefore dependent on engineering's plans for its own planning.

Companies must use the same kinds of rough-cut capacity planning tools for engineering resources as we previously described for manufacturing resources. This should be mandatory in engineered product environments, where the forecast and the product mix are constantly changing. The following discussion explores the pre–Sales & Operations Planning process for new products and engineered products.

New Product

Many companies will set up separate families for major new products even if they fit into one of the existing families. This is done to provide maximum visibility so that each major new product will be discussed at each Sales & Operations Planning meeting. The new product plans for engineering will obviously have to be coordinated very closely with sales plans, and any special requirements for new equipment, capacity changes, and materials availability must be made clear to manufacturing and purchasing so that the product is released on time.

New product introductions are universally difficult, largely because the unknown factors make them hard to plan, and very close monitoring is therefore required. Special attention during the Sales & Operations Planning process will help alleviate this problem by focusing on the critical issues involved.

The intent is not to cover all aspects of a new product introduction at the Sales & Operations Planning meeting; there will be regular detailed reviews of all aspects of the new product introduction. Rather, the emphasis is on how the new product affects aggregate planning, what impact it has on the business plan, and whether it poses resource conflicts with existing products.

Engineered Products

For engineered products, one must think of lead times and backlogs in much the same way as management considers them in manufacturing. You need to decide what your engineering backlog should be in order to provide good customer service. Classic input/output control can be an effective mechanism to accomplish this balancing act within the engineering department.

To see how input/output control can be effective in managing backlog, consider the case of a valve manufacturer that had one family where it engineered every product it sold. The company enjoyed an unexpected increase in business, 15 percent over forecast. In addition to this increase, the average engineering content of each customer order increased by 8 percent, and the department had a difficult time hiring and training enough engineers. Lead times, which marketing expected to be two weeks to get through engineering, were extended to four weeks, thereby causing major customer service problems as well as severe (and unfair) pressure on manufacturing. The company recognized the need for shortening the lead time, and a decision was made to hire engineers on a temporary basis. Using the input/output mechanism shown in Figure 4.4, the company was able to monitor and manage its growing engineering lead time and take appropriate action.

Figure 4.4 shows how input/output control works. The top section, marked "input," shows the planned input of hours to the engineering resource on the "plan in" line. The "actual in" line is the work that has arrived at the resource. The cumulative deviation over the last four weeks is +7 hours.

The output section shows on the "plan out" line the planned output. For the last four weeks, it was 115 hours per week. The actual output was less, thereby yielding a cumulative deviation of −25 hours.

The queue or backlog section shows on the "plan" line what the queue objective has been. The "actual" line shows what the combined input and output has yielded in terms of queue. At the current time, the queue has increased to 92, which is 32 hours over the planned queue. In order to get the queue down to the planned queue of 60, or a half a week's work, either the input would have to be reduced or the output increased. It might be necessary to work 15 hours of overtime each of the next two weeks in order to reduce the queue.

FINANCE'S PRE–SALES & OPERATIONS PLANNING MEETING PREPARATION

Finance must ensure that all the end-of-the-month reporting and processing is completed on schedule. It must also see that costs and pricing are up to date, as these will be used for all fiscal projections. In some cases, finance will be involved in making sure that data used by all departments are accurate and complete. Additionally, finance will convert all the projections into dollars and compare them against the business plan, reporting any major deviations that may require review or rethinking of management's strategy. In other words, finance should

Figure 4.4

Engineering										Input/Output Control		

Week	−4	−3	−2	−1	This Week	2/27	3/6	3/13	3/20	3/27	4/3	4/10

Input

	−4	−3	−2	−1	This Week	2/27	3/6	3/13	3/20	3/27	4/3	4/10
Plan In	105	105	105	115	115	115	115	115	115	115	115	115
Actual In	100	125	80	132								
Cum. Dev.	−5	+15	−10	+7								

Tolerance ± 20 Hours

Output

	−4	−3	−2	−1	This Week	2/27	3/6	3/13	3/20	3/27	4/3	4/10
Plan Out	115	115	115	115	115	115	115	115	115	115	115	115
Actual Out	120	115	100	100								
Cum. Dev.	+5	+5	−10	−25								

Tolerance ± 20 Hours

Queue

		−4	−3	−2	−1	This Week	2/27	3/6	3/13	3/20	3/27	4/3	4/10
Plan		80	70	60	60	60	60	60	60	60	60	60	60
Actual	90	70	80	60	92								
Dev.													

Tolerance ± 30 Hours

serve as a monetary watchdog group for the whole company. Finance typically benefits from an effective Sales & Operations Planning process, because good financial projections are dependent on the changes in the sales and production plans. As the quality of these plans improves, their conversion into financial plans will produce more accurate financial projections.

PREMEETING CONSENSUS

Some companies shape preliminary agreements before the Sales & Operations Planning meeting. This can be accomplished by conducting meetings between the various departments at the middle-management level. The advantage is that all alternatives can be thoroughly researched and spelled out, so that top management can most easily select the best ones when the formal meetings are held. The degree of premeeting activity depends on the complexity and size of the companies in question and the extent of the changes being suggested. In a small company where the general manager and his staff are closer to the front line, the Sales & Operations Planning issues can all be discussed in the regular meeting. In larger companies where management may be somewhat more removed, reaching preliminary consensus often helps management achieve the goals of the formal Sales & Operations Planning meeting in the shortest amount of time. In addition, preparation helps ensure that surprises in the Sales & Operations Planning meeting are eliminated.

One major international firm that has eleven factories and six major selling units must coordinate very closely to avoid harmful changes. Its manufacturing strategy is to maintain the flexibility to either coproduce or move production among plants in order to maintain employment stability. In addition, the company receives forecasted input from every selling division. Naturally, it takes longer for them each month to coordinate their Sales & Operations Planning process; this company holds its meeting on the fifteenth day of the month and does an excellent job.

A company's size, though, is not the only criterion for determining the value of preliminary meetings. As Thomas Connelly, vice president and assistant to the president of Hardinge Brothers, remarked, "Premeeting consensus became so important that we recently updated our Sales & Operations Planning policy document to include a specific list and timetable for pre-S&OP meetings. The goal is to ensure that monthly S&OP sessions will not be held up by lack of necessary data. It has caused us to hold back the timing of the main meeting until workday number sixteen, but we feel the results have been well worth the extra time required."

Obviously, premeeting consensus can be a critical factor in successful Sales & Operations Planning. Judge for yourself whether it could benefit your company, and tailor the process to meet the needs of your company.

REVIEWING THE FINANCIAL CONSEQUENCES OF CHANGES TO PLAN

The financial consequences of the changes initiated in the Sales & Operations Planning meeting are obviously critical. Ideally, it should be known prior to approval whether an operating decision will adversely impact the business plan. The only question is how the review of financial changes should be handled.

In some companies, the financial consequences are determined during the Sales & Operations Planning meeting. For example, if the new changes were discussed ahead of time, and finance was able to come to the meeting with their impact laid out, the general manager and his staff would see their effect in dollars before approving the changes. In other companies, there is a straightforward conversion from rates of output to dollars, so that their effects on the business plan of proposed changes are easily identifiable. Finally, there are a growing number of companies that have financial models tied into a terminal at the Sales & Operations Planning meeting. This enables the company to feed proposed changes into the model during the meeting and quickly make the consequences available to the staff.

One company has developed a financial model that mirrors the revenue section of its annual operating plan. Each month, the proposed Sales & Operations Planning data are loaded into the model, and the financial impact through the gross margin level is calculated. When changes are significant, they are highlighted at the meeting.

Often, however, it is not practical to discuss the changes to the operating plan and, prior to approving them, review their dollar impact as well. In these cases, two separate meetings should be held, each having a bearing on the other. One discusses what the goals are, while the other discusses what should be generated by executing the current operating plans. For companies just starting Sales & Operations Planning, assigning these two equally important issues to two different meetings may be best, as it reduces the scope of the challenge. Once you've demonstrated that you are capable of doing both reviews separately, it should be obvious whether they can be done together.

SUMMARY

- The more you can reduce the planning cycle, the quicker you can put a revised plan in action.
- Sales and marketing's primary responsibility in Sales & Operations Planning is to predict the future and to communicate any changes in plan to the rest of the company in a timely fashion.
- Whenever actual sales differ significantly from expected sales, sales and marketing must determine the causes. Understanding the reasons is important if you wish to know how to alter the sales plan.
- All apparent deviations in the marketplace must be "tested" to determine if they constitute real volume or mix changes rather than timing changes that do not represent long-term changes in demand.
- Revisit your assumptions and vulnerabilities each month. Were you right in your prediction about how economic factors would affect your sales? If not, readjust your plans.
- Manufacturing and engineering should be able to perform rough-cut capacity planning to determine whether the company can cost-effectively adjust capacity to accommodate changes in the sales plan.
- Premeeting consensus can greatly streamline the formal Sales & Operations Planning and allow participants to focus on management issues.

Chapter 5
Preparing the Data

Music creates order out of chaos.
(Yehudi Menuhin, American musician)

Sales & Operations Planning is a process of evaluating data and making plans, then communicating that information throughout the company. Even though you are dealing at the aggregate level, the data burden is nevertheless significant. When organized properly, the format and choice of numbers will make it easier to make decisions, instead of serving as obstacles that require valuable time and energy to decipher.

Well-organized data not only enable the Sales & Operations Planning team to determine what needs to be done to support the numbers, but provide a rock-solid foundation for carrying out "what-if" analyses and exploring alternatives.

In addition to the question of what information will be conveyed, other key data issues include cutoff, timeliness, and formatting, or presentation. We'll talk about each one and then provide guidelines for optimal data presentation, including a sample format.

Whenever the presentation makes the information easy to grasp, whenever the consequences are obvious, and whenever alternatives are quickly available, decision-making is enhanced. "Make the data transparent" was how Ollie Wight described the goal. That means ensuring that the users can "see through" the data to understand what they mean and how they were derived.

Reports that are incomplete, cumbersome to use, and filled with complex calculations make it difficult for people to accept, question, or reject the information. Blindly following calculations, hoping that they are leading you in the right direction, is no way to run a business. On the other hand, having to turn away from the figures because of uncertainty forces you to rely completely on "gut feel."

PERSPECTIVE

There are three major activities involved in providing good information to the Sales & Operations Planning process. The easiest of the three is presenting it in a helpful manner. A more difficult task is to ensure that it is both accurate and timely. Still more challenging is using the data to arrive at important company decisions.

For example, good data will assist the person responsible for forecasting. Analyzing these data and then adding to them all of the non-quantifiable factors to generate a new prediction requires talent and courage. Everyone knows that it will be an inaccurate guess, and everyone realizes that it will have a major impact on the future actions of the company. Yet, not providing the figures to form the basis for the Sales & Operations Plan is a guarantee that the company will be less competitive.

Agreeing on targets for each family of products—the desired ending inventory for make-to-stock products and the desired ending backlog for make-to-order products—is another area of controversy. You need to establish these targets in order to determine your course of action. Therefore, they are inescapable.

Still, one person's judgment as to what they should be is apt to differ from another's. Generally, marketing will lobby for more finished goods, recognizing that it will have a favorable impact on customer orders since it increases the odds that shipments can be made despite inaccurate forecasts. Finance, on the other hand, will be quick to point out the costs associated with having extra inventory, and will generally be pushing in the opposite direction. Manufacturing may be quiet during this phase of the discussion, but will speak up when asked to make fast changes to its output rates. Typically, manufacturing will want more time, while other members of the general manager's staff will press hard to accomplish it in less time.

Presenting pertinent, accurate data in a readable manner doesn't make the decisions right, but it certainly makes the decision-making process immeasurably easier.

CUTOFF

All departments must agree about the data cutoff dates. Sometimes there are differing opinions, because sales may prefer to work on a calendar month, while accounting would like to work on a 4, 4, 5 quarterly calendar. Manufacturing may prefer to work with periods of equal size, so that rates per period make more sense. Without a common cutoff, the exchange of information creates confusion rather than additional knowledge. Financial reporting requirements, though, usually prevail.

One Midwestern manufacturer of automotive parts creatively solved the problem of different-sized periods by converting the presentation of the data from a calendar basis to a daily rate basis. The key to the success in this case was that even though the months were of different size, all the data were presented in terms of daily rates (sales and production). As a result, visibility of change was greatly enhanced. The following example illustrates how their system works.

	April	May	June
Selling Days:	19	20	25
Manufacturing Days:	17	20	25
Sales Plan:	19,000	20,000	25,000
Sales/Day:	1,000	1,000	1,000
Production Plan:	17,000	20,000	25,000
Production/Day:	1,000	1,000	1,000

If you looked only at the sales plan and production plan numbers, you might assume that both are increasing in May. In fact, they are increasing in volume, but not in rate. The rate figure therefore gives you a much better idea of where real change is occurring. Also notice that there are different numbers of sales and production days in April. If these vary within a month, the difference should be very visible. Accounting was pleased because no one fought the calendar that they wanted to use. As the example shows, if everyone can't agree on cut-offs, a conversion to rates can help.

TIMELINESS

To be effective, information must be up to date. If the month ends on the thirtieth, then the data processing systems should be designed to provide monthly summaries as soon as possible. Any delays will be an impediment to the Sales & Operations Planning process, as they will add to the lead time of decision-making and responding to change.

Be careful about accounting for every last piece of data in order to do effective Sales & Operations Planning. Sales & Operations Planning is an aggregate approach, and therefore not very sensitive to a few small missing pieces. It would be a false sense of security to think that a precise starting point will significantly alter your long-range ending point. In fact, the extra time lost by waiting to gather the missing data may cost you a timely decision. Better to have the vast majority of information available in a timely meeting than 100 percent of the data in a delayed meeting.

DATA PRESENTATION TOOLS

In order to carry out Sales & Operations Planning, you must present data so that three types of information are readily visible:

- What was our past performance?
- Where are we today?
- What are the proposed or current plans for the future?

This sounds simple enough, but the fact is, software is just emerging that readily pulls together data from all departments into a single report answering all three questions. Since most dedicated mainframe software packages aren't set up to perform the necessary data integration, many companies have turned to PCs and popular spreadsheet programs, such as Lotus 1-2-3 and Supercalc. A number of dedicated data consolidation and presentation packages are now on the market.

All of the data needed to make a decision about each family must be readily available and on a single report if possible, so that any one of the participants in the Sales & Operations Planning meeting can understand all of the information about the family in terms of bookings, production, inventory, backlog, and shipments. This will also prevent having to hunt for information.

The need for properly consolidated data is well illustrated by the case of one company that had just begun Sales & Operations Planning. The general manager wanted to know why the company did not achieve the plan, and where the planning team thought the market was going. Since the existing reports only projected future numbers, he had to ask repeatedly what had happened in the previous months. Each time he would ask, everyone would have to shuffle through their notebooks to find the necessary reports and printouts from the previous months. At the end of three hours, the meeting adjourned with little accomplishment and high frustration.

The general manager did, however, suggest that a proper method of data consolidation and presentation be implemented before the next meeting took place. This was accomplished with a personal computer and a concerted spreadsheet programming effort. The following month, the Sales & Operations Planning meeting took a different turn. All necessary data were readily available, so that rather than stumble over the numbers, the Sales & Operations Planning team was able to make decisions based on them.

Figure 5.1 shows a sample report for a family. There is no one standard presentation that is going to work for all companies. The sample format, though, presents the information essential to most companies in a straightforward and easy-to-read fashion. Some companies will want the information displayed in dollars, others will want it in units, and still others will want to see capacity information. Some

Figure 5.1

TODAY →

MONTH	J	F	M	A	M	J	J	A	S	O	N	D	Yearly Total	J	F	M
Days in Month	20	20	20	20	20	20	20	20	20	20	20	20	240	20	20	20
SALES PLAN																
Last Year's Actuals																
Annual Business Plan																
Current																
Proposed																
Deviation																
Cumulative Deviation																
PRODUCTION PLAN																
Annual Business Plan																
Current																
Proposed																
Deviation																
Cumulative Deviation																

Figure 5.1 *(continued)*

INVENTORY														
Annual Business Plan														
Current														
Proposed														
Deviation														
Days Cover														
BACKLOG														
Annual Business Plan														
Current														
Proposed														
Deviation														
SHIPMENTS														
Annual Business Plan														
Current														
Proposed														
Deviation														
Cumulative Deviation														

companies will prefer to present dollars and units on separate pages, while some want it all on the same page. Clearly, there are many styles that different companies can choose from. For purposes of ease of understanding, the data in Figures 5.2 and 5.3 are shown in units.

The reports are divided into five sections:

1. Sales plan
2. Production plan
3. Inventory
4. Backlog
5. Shipments

The first segment of the sample format deals with the sales plan. To the left of the line labeled "today" is space available for historical information. A number of companies may want to show more than the three preceding months included on this report; we've seen as many as twelve months displayed. To the right of "today" is planning information for the future, extending through the selected planning horizon. In this example, the next twelve months are shown. Because executives like to review the status of the fiscal year, we have also incorporated a "yearly" total. In this example, the calendar year and the fiscal year are identical.

Let's step through each of the rows. "Last year actuals" reflect customer orders booked in the previous year. "Annual business plan" contains the expected bookings for this fiscal year. "Current" is the approved sales plan from the prior Sales & Operations Planning meeting, including any final adjustments made to it. "Proposed" represents what the sales and marketing groups are projecting based on their most recent evaluation of the marketplace. The first three columns offer an opportunity to show actual bookings for the last three months. "Deviation" is the difference between actual bookings and the annual business plan. "Cumulative deviation" is the running total of this same difference. These last two rows would only contain figures to the left of "today," as they are hindsight calculations.

The second segment of the report contains information regarding manufacturing's production plan. As with the first segment, to the left of "today" is information covering the prior three months. "Annual business plan'' shows what was expected to be built (i.e., completed products) for the business year, of which three months have gone by and nine months remain. The ''current'' row presents the approved plan from last month's meeting; the ''proposed'' row contains the monthly output suggested by manufacturing for the next twelve months. To the left of ''today'' is where the actual output for the past three months could be displayed. The ''deviation'' row reflects the difference between actual output and the original commitment made for the annual business plan. The ''cumulative deviation'' row is the running summary of the differences. Last year's annual business plan and last year's actual output are not displayed but certainly could be added.

The third segment contains the inventory information. ''Annual business plan'' shows the expected month-end finished-goods inventory for each month. The ''current'' row shows the inventory targets as approved in last month's meeting. The ''proposed'' row shows the projected month-end finished goods as determined by taking the starting figure (shown in the column immediately to the left of the ''today'' line), adding the proposed production plan, and subtracting the proposed sales plan.[1] ''Deviation'' represents the difference between the actual finished-goods inventory and the annual business plan figures.

''Days cover'' expresses the size of the finished-goods inventory as a measure of time. For the preceding three months, this would be calculated by taking the actual inventory and dividing it by the following months' daily sales rate (actual sales divided by number of workdays). For the future months, the figure is derived by converting the proposed inventory into the number of days it represents in terms of the next month's proposed sales.

The fourth segment displays backlog information both for the past three months as well as for the future. In a make-to-stock environment, this is really back orders. In make-to-stock where end products are being shipped off the shelf, the expectation is that bookings and shipments equal one another. For a make-to-order product, however, this is seldom the case—a significant number of the orders being booked this month will not be shipped in the same month and could stretch out for a considerable number of months into the future. Therefore, we need to convert bookings to backlog. This will be covered later.

[1] This resulting figure should be adjusted by the net change in backlog discussed in the formula segment.

Figure 5.2

TODAY →

MONTH	J	F	M	A	M	J	J	A	S	O	N	D	Yearly Total	J	F	M
Days in Month	20	20	20	20	20	20	20	20	20	20	20	20	240	20	20	20
SALES PLAN																
Last Year's Actuals	154	156	175	172	191	198	190	175	174	162	155	136	2038	–	–	–
Annual Business Plan	160	160	180	180	200	200	200	180	180	160	160	140	2100	–	–	–
Current			180	180	200	200	200	180	180	160	160	140	2110	160	160	–
Proposed	164	166	187	190	205	205	205	185	185	160	160	150	2162	170	170	190
Deviation	+4	+6	+7													
Cumulative Deviation	+4	+10	+17													
PRODUCTION PLAN																
Annual Business Plan	170	170	170	170	200	200	200	200	160	160	160	160	2120			
Current		170	170	180	200	200	200	200	160	160	160	160	2119	170	170	–
Proposed	164	165	168	180	200	210	210	200	200	160	160	160	2177	180	180	180
Deviation	–6	–5	–2													
Cumulative Deviation	–6	–11	–13													

Figure 5.2 *(continued)*

INVENTORY																	
Annual Business Plan	161	171	181	171	161	161	161	161	181	161	161	161	181	—	185	195	—
Current				155	155	155	155	155	175	155	155	155	175	—	185	195	—
Proposed		163	165	155	150	147	142	137	152	167	167	167	177	—	187	197	187
Deviation		-8	-16	-16													
Days Cover		20	18.3	16.3	14.6	14.3	13.8	14.8	16.4	20.8	20.8	23.8	22.1	—	20.8	22	19.7
BACKLOG																	
Annual Business Plan		10	10	10	10	10	10	10	10	10	10	10	10		15	15	
Current			15	15	15	15	15	15	15	15	15	15	15		15	15	
Proposed		12	15	24	29	29	31	21	11	11	11	11	11		11	11	11
Deviation		+2	+5	+14													
SHIPMENTS																	
Annual Business Plan		160	160	180	180	180	200	200	180	180	160	160	140	2100	160	160	
Current				180	180	200	200	200	180	180	160	160	140	2105	160	160	
Proposed		162	163	178	185	203	215	215	185	185	160	160	150	2161	170	170	190
Deviation		+2	+3	-2													
Cumulative Deviation		+2	+5	+3													

Referring to segment five, shipments, the "annual business plan" reflects what was expected to be shipped for the fiscal year. "Current" is last month's approved shipment plan. "Proposed" contains what is expected to be shipped based on the proposed sales plan and production plan. "Deviation" represents the difference between actual shipments and the annual business plan. "Cumulative deviation" is the running sum of this difference. Observe that the number of workdays is listed for each month. Earlier in the chapter we explained how one could calculate a per-day rate using the number of workdays divided into the period quantity. Even if you don't do this it is essential to know the number of workdays to better understand the data. You are going to have more volume in a month with more workdays.

Using the Report

Having described the information that would be presented in the report, let's insert some data as shown in Figure 5.2 and review how it might be used. This year's annual sales plan projects a 3 percent increase. Last year, we sold 2,038 units, and this year we're forecasting 2,100 units. Reviewing the previous sales plan, you'll note that the monthly figures are identical to those on the annual business plan and yet the yearly total reflects 10 units more. This is not a mistake, but rather a judgment from the sales department regarding the increased actual sales in January (+4) and February (+6). This total of 10 units more than the sales plan was classified as "abnormal" demand and as such is considered plus business, hence the addition to the yearly figure. It was not interpreted as a trend and thus did not alter the future predictions. On the other hand, the +7 deviation in the past month has caused the sales department to revise its projection of sales. The proposed sales row has been increased by a modest 2 percent over the original business plan.

The production plan segment reveals that in January and February manufacturing fell short of their commitment by a total of 11 units. As a result, the production plan for April has been increased by 10 units to make up most of the shortfall. Manufacturings's response to the proposed sales plan is not to change their prior commitment for April and May, but to increase output by 10 units in both June and July.

Furthermore, manufacturing plans a 40-unit increase in output in September to compensate for the additional increase in the sales plan.

The net effect of what took place in January and February, when actual sales exceeded forecast by 10 units and actual production fell behind expected production by 11 units, is a variation of 21 units. For the numbers to reconcile properly, this variation should show up in the inventory segment. Note that actual inventory at the end of February was 16 units below plan. The missing 5 are back orders and are displayed in the backlog segment.

Other Formats

Figure 5.3 shows a format similar to the one in Figure 5.2. It contains some additional lines of data that would be appropriate for a make-to-order format. In the segment that presents production information, a new row would be called "uncommitted production" or "available-to-promise." In the backlog section the open customer orders are shown by promise date. The uncommitted production would be calculated by comparing the proposed production to the open customer orders by promise date in order to reflect what amount of each family's production plan is available for future promising to customers.

In a make-to-stock company, a bookings forecast is appropriate, because bookings and shipments usually occur in the same time period. In a make-to-order company, however, manufacturing is not really interested in when the customer places the order. What manufacturing really needs to know is, "When does the customer need the product?" Therefore, a forecast of demand is required. If the sales plan is still a bookings forecast, then to reflect the difference between bookings and demand, companies may develop ratios referred to as "book to ship" or "book to demand." For example, let us assume 100 orders arrive this month, and based on our historical performance we would expect to ship 20 percent in the same month, 40 percent next month, 30 percent the following month, and the remaining 10 percent in month four. If sales and marketing project bookings, this ratio would permit the company to convert their forecast of incoming orders into a projection of demand.

The formats that we have presented will support make-to-stock, make-

Figure 5.3

TODAY →

MONTH	J	F	M	A	M	J	J	A	S	O	N	D	Yearly Total	J	F	M
Days in Month	20	20	20	20	20	20	20	20	20	20	20	20	240	20	20	20
SALES PLAN																
Last Year's Actuals	154	156	175	172	191	198	190	175	174	162	155	136	2038			
Annual Business Plan	160	160	180	180	200	200	200	180	180	160	160	140	2100	—	—	—
Current			180	180	200	200	200	180	180	160	160	140	2110	160	160	—
Proposed	164	166	187	190	205	205	205	185	185	160	160	150	2162	170	170	190
Deviation	+4	+6	+7													
Cumulative Deviation	+4	+10	+17													
PRODUCTION PLAN																
Annual Business Plan	170	170	170	170	200	200	200	200	160	160	160	160	2120	170	170	—
Current			170	180	200	200	200	200	160	160	160	160	2119	170	170	180
Proposed	164	165	168	180	200	210	210	200	200	160	160	160	2177	180	180	180
Uncommitted Production				166	190											
Deviation	-6	-5	-2													
Cumulative Deviation	-6	-11	-13													

Figure 5.3 *(continued)*

	1	2	3	4	5	6	7	8	9	10	11	12	13	Total	14	15	16
INVENTORY																	
Annual Business Plan	161	171	181	171	161	161	161	181	161	161	161	161	181				
Current		163	165	155	155	150	147	142	137	152	167	167	167	177	—	185	195
Proposed	163	165	155	150	147	142	137	152	167	167	167	177	177	—	187	197	187
Deviation	−8	−16	−16														
Days Cover	20	18.3	16.3	14.6	14.3	13.8	14.8	16.4	20.8	20.8	20.8	23.8	22.1	—	20.8	22	19.7
BACKLOG																	
Annual Business Plan	10	10	10	10	10	10	10	10	10	10	10	10	10				
Current		15	15	15	15	15	15	15	15	15	15	15	15			15	15
Current Booked Backlog		14	10														
Proposed	12	15	24	29	31	21	11	11	11	11	11	11	11		11	11	11
Deviation	+2	+5	+14														
SHIPMENTS																	
Annual Business Plan	160	160	180	180	200	200	200	180	180	160	160	160	140	2100	170	170	190
Current			180	180	200	200	200	180	180	160	160	160	140	2105			
Proposed	162	163	178	185	203	215	215	185	185	160	160	160	150	2161	170	170	190
Deviation	+2	+3	−2														
Cumulative Deviation	+2	+5	+3														

to-order, or a combination of both. In an engineer-to-order business, it would be very important to add a section reflecting backlog in engineering. Such additions may appear obvious and most of them are. Be careful of overwhelming the readers, however; we have seen companies overly creative in format and overzealous in terms of displaying the same information in a variety of ways, in the hopes that it will help. In the end, the efforts simply made the review very inefficient. The format that we are presenting should stimulate you to think about what is the minimum amount of data necessary and how it might be displayed for maximum visibility and quick understanding.

Use of Graphs

While columnar data, as presented in a spreadsheet, are essential, many companies are finding that the use of graphs can add a new dimension to Sales & Operations Planning data. Graphs can point out interrelationships in ways that columnar data can't. Moreover, graphs show the impact of data in ways that everyone can readily appreciate, because they are fast and easy to read.

SALES & OPERATIONS PLANNING FORMULAS

There are two basic formulas that are useful for calculating and understanding what manufacturing must produce to hit the agreed-on targets, inventory, or backlog. One applies to families that are make-to-stock, the other to families that are make-to-order.

For make-to-stock products, the formula is as follows:

PRODUCTION PLAN = FORECAST + (DESIRED ENDING
FINISHED GOODS − STARTING FINISHED GOODS)

Marketing would supply the forecasts covering the period of time being reviewed. The change in levels of finished goods reflects the difference between your starting point and the approved target for the ending of the period under consideration.

Let's say that marketing expects to sell 1,000 units in the next twelve months, and the general manager has approved raising the finished-goods level from 100, where it is today, to 200 by the end of this twelve-month horizon. Total output—that is, the production plan—will have to come to 1,100 units to accomplish this. The relationship is a very simple and straightforward one. If you wish to build finished-goods inventory, then output from the factory must exceed incoming orders, or else inventory won't increase. On the other hand, if you wish to decrease finished goods, output from the factory must be less than the incoming orders.

To implement a change in output rate, it may not be possible for the factory to start a revised rate of production immediately. (Chapter 2's discussion of time fences covered this issue.) In the above example, let's assume there are 220 workdays available in the next twelve months, but the factory cannot increase its current level of output for the first 90 days. The entire change must then be spread over the remaining 130 days.

Frequently, it is desirable to express finished-goods inventory in terms of time, i.e., in months' or days' coverage. To see the total inventory going up or down may not be as significant as translating it in terms of days' worth of inventory. This can be accomplished by dividing the inventory at a given point in time by the subsequent period's forecast of sales. A number of companies call this "forward coverage," and are more comfortable using it when exchanging their opinions of what the desired level of inventory should be.

For make-to-order families, the formula is as follows:

$$\text{PRODUCTION PLAN} = \text{FORECAST} + (\text{STARTING BACKLOG} - \text{DESIRED ENDING BACKLOG})$$

Again, marketing must supply the forecast of what new orders are expected during the period under consideration. The general manager must approve the change in the backlog in terms of whether it should be set higher or lower than where it is at the beginning point.

To decrease the backlog, production must exceed incoming orders, or else the backlog will not decrease. On the other hand, if the goal is to increase the backlog, the output from the factory must be less than incoming order rates.

SUMMARY

- Pertinent, accurate, timely data presented in a transparent way allow the general manager and his staff to concentrate all of their energies where they should be—in running the business more effectively.
- No one should trip over the data because they are poorly organized or presented.
- Cutoffs for data processing are critical for making Sales & Operations Planning work. Each department must agree to workable dates and compromise when necessary to establish a common one for the company.
- Good data are timely data. Make sure reports are distributed as soon as they are generated and that planning meetings are scheduled to occur quickly thereafter.
- Data must be consolidated to show past performance, current position, and future plans.
- Keep Sales & Operations Planning reports simple, and use graphs in addition whenever possible.
- Be prepared to develop the tools to support your reporting needs.

Chapter 6
Operating the Process

Conducting is a real sport. You can never guarantee what the results are going to be, so there's always an element of chance. That's what keeps it exciting.

(Aaron Copland, American conductor)

The Sales & Operations Planning meeting is not only a means for agreeing on and updating plans, it's also a key opportunity to enhance the overall communication process throughout the company. This is important because companies typically don't communicate as well "horizontally," across departments, as they do "vertically," within departments. There are several reasons for this imbalance. First, companies often don't see the need for interdepartmental communications. "Why does someone in manufacturing need this information?" someone in sales might ask. Second, people are too busy—the pressures of everyday business are such that people simply don't take the time to work on communicating with other departments. Finally, there's a turf issue: "This is *my* area. Why should I share or have to defend the information?"

In most companies, the various departments have very good intentions to communicate well and really believe they are doing so; however, in reality, the vehicle (Sales & Operations Planning) just has not been set up. Good information that is not used in a timely manner ages badly, and the consequences of delayed decisions are likely to hurt the overall performance of the company.

Of course, all of these excuses are impediments to running the company, and must be eliminated for Sales & Operations Planning to be effective. Once Sales & Operations Planning is in place, people will realize the value of interdepartmental communication, and will see the benefits for their own departments as well as for the company as a whole. The major benefit for the business is that effective communication helps develop objectives and facilitates making decisions. It also eliminates surprises. And the fewer surprises, the more likely management will be able to meet its goals without resorting to crisis maneuvers.

Another aspect of the communications issue is the keeping and distribution of a set of minutes recorded during the meeting. The minutes are very important, because if you ask ten participants to state what went on in the meeting two days later, chances are you'll get ten different versions. Memory is too subjective and fleeting. Put the decisions on paper as they are made.

As for distribution, too many companies regard the minutes as "eyes-only" documents for the Sales & Operations Planning participants. Limiting distribution only limits the effectiveness of the management process, because 99 percent of the company remains in the dark about the decisions made behind closed doors. The minutes of the meeting should be disseminated throughout the company, as they represent an excellent opportunity to include everyone on the "company team."

In addition to improving communication, the Sales & Operations Planning meeting provides a unique opportunity to transform the all-too-pervasive problem mind-set, which looks backward and emphasizes who did what to whom, into a "solution" mind-set, which looks forward and concentrates on positive responses. Without a valid plan, problems tend to consume the bulk of the discussion: "Why didn't we get the expected results?" . . . "What went wrong?" . . . etc. A well-conceived plan not only makes problems easier to analyze, it also permits the general manager to ask important questions in the other direction: "How do we hit our target next month?" . . . "What is the competition doing, and what should we do in response?"

Of course, it's necessary to find answers to problems so corrective actions can be taken. But consider what we could accomplish through an alternative approach, looking at situations from a standpoint of success. This means asking questions like "What enabled us to carry out the plan?" and "Why did we hit the numbers?" This orientation provides a richer set of insights into the way your company works. Besides, success breeds success, and by looking at why things actually work, people begin to view the world in terms of positive opportunities rather than crises waiting to detonate. In short, Sales & Operations Planning meetings offer people a whole new way to think about their jobs and the way the business is run.

Before actually sitting down to run a meeting it's important to consider several general issues that determine how effective the meeting will be:

1. *Plan out a year's worth of meetings in advance.* Companies that are most successful with Sales & Operations Planning lay out dates for all key events, including strategic planning, annual budget planning, and monthly Sales & Operations Planning meetings. If you expect people to attend the meetings, they must be able to plan ahead accordingly. You can therefore minimize missed meetings by maintaining the long-range calendar.

2. *Distribute a Sales & Operations Planning "kit" before the meeting.* Every company should develop its own meeting "kit" containing the following information:

- *An agenda.* This is a key element. Specific agenda topics are addressed below. Remember, while having an agenda is important, sticking to it is equally important. Given the complexities of running a manufacturing company, it's easy for a planning meeting to drift off on a tangential current. It's very tempting for each staff member to bring the latest "crisis of the hour" with him. No doubt, marketing would be very anxious to talk about a major customer's difficulty, whereas manufacturing may be encountering a serious production problem, while engineering is confronted with a tough design issue. The general manager must always be alert for signs of such navigational problems, and steer the ship back on the designated course if the discussion doesn't appear to be generating productive results.

- *All of the premeeting plans created by the various departments.* During the premeeting process, departmental plans were developed and shared. Whatever revisions and compromises were made between the departments should be reflected in the individual plans. If all attendees review the other department recommendations prior to the meeting, they will come prepared. Problems can then be quickly addressed, reasonable alternatives developed, and decisive plans for the future approved.

- *Analysis of the assumptions and vulnerabilities.* A critical part of the Sales & Operations Planning process is to state the assumptions by which each department created its plans. All participants will need to understand these assumptions if they are to understand the basis of each other's plans.

3. *Block out enough time for each meeting.* Be realistic. If your company is just starting Sales & Operations Planning meetings, the sessions may initially be lengthy. Take heart in the fact, though, that many companies significantly reduce the time required to carry out Sales & Operations Planning meetings as they become more proficient at the mechanics and refine their premeeting homework. Even companies that have mastered the art of Sales & Operations Planning, however, should expect to have longer meetings during times of abnormal change. While we can't specify the precise amount of time you'll need, just bear in mind that all participants must allocate sufficient time for each meeting, and should have realistic expectations about the amount of effort that will be required in the early stages.

4. *Create a positive environment for the meeting.* The attitude the company takes toward Sales & Operations Planning is vital to the success of the process. The general manager has the responsibility of ensuring that the Sales & Operations Planning process is perceived in a positive way. Focus on the future and action plans instead of dwelling on past problems. It is also important that the meetings be taken seriously. The general manager should insist that the meetings start on time (some companies actually lock the door the minute the meeting starts!) and that, with the exception of fire alarms, there be no interruptions—*especially* no telephone calls.

While the above measures do not guarantee a successful meeting, their absence is a sure road to failure. With the preliminaries out of the

Figure 6.1

---

1. Special issues
2. Company performance review
   a. customer service
   b. financial and business
   c. department-by-department (sales and marketing, engineering, and manufacturing) review in aggregate
3. Reviewing assumptions and vulnerabilities
4. Family-by-family review
5. New product discussion
6. Special projects
7. Review of meeting decisions
8. Critique of process

---

way, we can now turn to the actual flow of the meeting, which will cover the topics listed in Figure 6.1

During the meeting the general manager and the attendees should engage in a dialogue in which the participants present their plans and alternatives. Note that this is in the form of sharing information, rather than an interrogation by the general manager. The exchange allows the general manager to ''bullet-proof'' the plan and make sure that the underlying numbers and logic are sound. Here's how the process unfolds.

1. OPENING MOVES: RAISING SPECIAL ISSUES

The general manager should make a ''kickoff'' statement, bringing to the floor any overriding issues of special significance. These issues set the tone for the whole meeting. Perhaps the general manager has a concern about a decline in current customer service. Maybe he needs to convey certain financial pressures from above that affect short-term performance. Maybe he has noticed an economic trend that could affect the company. In a large corporation, the general manager might want to shed some light on what corporate management expects from the division in terms of cash flow and profit, and how these expectations might have an effect on the division's strategic thinking. There are those times when the total company is struggling and your division is asked to help pick up the slack.

The "kickoff" should be an opportunity for anyone to raise points that might be helpful for a *positive* understanding of the business. We stress "positive" because the liability of opening the session with special issues of concern is that "doom and gloomers" may take the opportunity to gripe about all the "can't do's," and pollute the meeting before it even starts. The general manager must therefore make it clear that the purpose of the kickoff is to promote a better understanding of special issues, not to air special complaints.

2. PERFORMANCE REVIEW

Performance measurements are not ends in themselves; rather, they are used to take the pulse of the business and catch problems before they become crises. Moreover, by reviewing performance, you should be able to uncover the underlying causes of any difficulties and find appropriate remedies.

Let's say you're shipping 85 percent of the part numbers ordered by your customers. That means you have stock on hand only 85 percent of the time. The source of the problem could be inaccuracies in your forecasts, which would mean you have a marketing problem, or it could be a quality/scheduling/execution difficulty. Regardless, the problem must be further researched to determine what corrective actions need to be taken.

Performance data, whether in tabular or graphic form, should be applied to the following areas:

Customer Service

The lifeblood of your business is satisfying your customers. Step one in any performance evaluation, therefore, is determining how well you're fulfilling customer needs.

Customer service measurements will differ for make-to-stock and make-to-order products. The following items are the major performance measurements for each type of product:

1. Product availability—for make-to-stock. Calculated by units shipped versus units ordered, or line items shipped versus line items ordered.
2. Delivery to promise—for make-to-order and any make-to-stock back orders. Calculated by number of customer orders shipped this week versus total orders promised.
3. Delivery to customer request date for make-to-order. Compares gaps between the date requested and the date shipped.
4. Current delivery lead time versus your goal, by family for make-to-order. Today's backlog expressed in time.
5. Past due or back order expressed by age for make-to-order and make-to-stock.
6. Quality (all products). Measured in terms of scrap and/or field complaints.
7. Customer returns (all products).

A key issue in customer performance is making sure that you know how your customers and the marketplace really evaluate your service—not just how *you* think they should evaluate your service. To determine customer perceptions of your company, your marketing department should consider conducting a special survey, or you may hire an outside group to do the survey for you. The latter approach may entail some expense, but it could get you the insights that only an outsider can provide. Either way, you must get an *objective* measurement of how the marketplace perceives your services.

Financial Performance

A review of overall financial performance with comparison to budget is necessary before delving into a more detailed, family-by-family analysis. It is important to get a broad sense of how you are doing with respect to your objectives. Therefore at all times it is essential to know the original budget or plan for the year, any revised plan that has been approved, and how the financial picture stacks up to both the original and revised plans.

Items to be reviewed for financial performance include:

1. Bookings (orders received): dollar value of new orders received
2. Shipments: dollar value of actual shipments
3. Backlog: dollar value of orders received but not yet shipped (aged if appropriate)
4. Inventory (finished goods, work in process, and raw materials): the value of those categories at cost
5. Profit: the bottom line
6. Cash flow: the difference between dollars coming in and dollars flowing out.

Department-by-Department Review in Aggregate

The objective of these aggregate performance reviews is to get an overall sense of how each department is performing prior to moving to the family level. The aggregate review yields very important information. Where performance to the plan is within the preestablished tolerances, it is safe to conclude that things are in good shape and under control. Whenever the aggregate performance is outside the acceptable tolerances, the opposite is true. Almost always, more words are appropriate to explain why this has occurred. These words typically are designed to give a "feel for the situation," as the detailed explanations often vary by family. There are occasions, however, where the variations are across the board, and can be explained at this point in the meeting.

The aggregate review of each department prior to the family-by-family review enables the general manager and his staff to look for opportunities to correct any problems, to seize new opportunities if they are available, or to reallocate resources among families if that will help.

Typical reviews in aggregate might include:

1. Sales Plan

How have you done in the most recent period? Were the total incoming orders in line with your forecasts? Are the year-to-date actual sales within established tolerances? If there have been deviations up or down, do you expect them to continue? There are times when both the reasons for the differences and the plans for addressing them are better handled in the family-by-family review.

Having summarized where you are, you need to discuss what has occurred, and what the future looks like, especially in relation to prior forecasts. You can learn from the past, but agonizing over it is unproductive. Looking forward to describe the shape and needs of the marketplace, based on where you are and what you have learned, is another responsibility of the sales and marketing group.

If there are any major programs or promotions under way, sales and marketing will likely report on their status. Certainly these could include hiring new people, training new and experienced people, new advertising programs, etc. This whole dialogue should be an open, positive discussion designed to provide all participants in the Sales & Operations Planning process with a good sense of the marketplace and how the company is faring in it.

2. *Manufacturing Plan*

Like his counterpart in sales, the manufacturing manager should review the immediate past period in terms of how well the plan has been achieved. All of the support departments are walking a tightrope—achieving the planned output while keeping costs to a minimum requires a delicate balance. Tremendous pressure is exerted on production people to increase the output to a higher rate while lowering unit costs, the difference between the two being the vital productivity gain. The production people take great pride in being "can-do" people and as a result may set very difficult improvement goals for themselves. Attaining productivity improvements requires doing things differently—and doing things differently does not always work out as successfully or as quickly as planned.

As with the sales performance review, we should look at a period performance as well as a longer span of time, perhaps year-to-date, to summarize performances. Where significant differences have occurred, an explanation and planned corrective action should be forthcoming. The manufacturing manager will represent manufacturing and its ability to support future plans. The plans *will* be supported; the question is *when* and *how*.

3. *Engineering Plan*

Engineering can be critical to the Sales & Operations Planning process. If you are an engineer-to-order company, engineering must address the sales plan before manufacturing does; if engineering does not have the resources to support the plan, then it is academic whether manufacturing can support it. You might also need to review the impact on the lead time of processing an order through the engineering cycle.

All companies need to be concerned about providing sufficient engineering support for new products, as they are essential for future growth. We will discuss new products in more detail later in this chapter.

3. Reviewing Assumptions

It's important to review the assumptions that have been previously made about the market, the economy, the competition, and internal factors within your company. The goal is to revalidate the assumptions that are true, to identify the ones that are no longer correct, and to document additional ones. The points to review are:

1. Has the marketplace changed? If so, how?
2. Have economic predictions been borne out?
3. Have there been any changes in the competitive situation?
4. Are there any changes in your company that will affect business?
5. Have your exposures and vulnerabilities changed?
6. Are there any new assumptions that haven't been considered before?

There are, of course, those sets of assumptions that apply to the total business. There are others, though, that apply to specific families and would best be left to the family-by-family review, which is next on the agenda.

4. Family-by-Family Review

This phase of the meeting is designed to lead to approval of the plans (sales, manufacturing, engineering, and finance) for each family. The process entails looking at the data presentation for each family, investigating the demand side (sales and marketing), and then getting the other departments' analysis of the new proposals, so that the group as a whole can reach a decision. It's important to walk all the way through

a family and gain consensus on it before moving on to the next one.
Three questions need to be asked on the demand side for each family:

1. *History: how did you do in the past?* The answer is obtained by conducting an honest evaluation of what has transpired during the past period. This means reviewing sales performance, with sales and marketing explaining why sales have been higher, lower, or equal to the forecast. You need to determine the reason for the sales performance. Is it something you or your competitors are doing? Is the economy driving the change? Is it the marketplace that has changed?

It is also important to ask, "Have we met our customer service objectives for this family?" If customer service is not being achieved, then you need to determine why without jumping to conclusions. Is the problem poor operations performance? Poor promising tools? Or poor forecast accuracy?

2. *Current status: how are you positioned today?* In a make-to-stock family, it's necessary to ask, "Do we have the finished-goods inventory to be positioned in the marketplace?" In a make-to-order environment, the key question becomes, "Is our current delivery lead time where we would like it to be?" For each family, you need to reevaluate your inventory and lead-time objectives.

3. *Future planning: where are you going?* After reviewing past performance and the company's current position and competitive strategy, you are prepared to revise your projections. What will be the future demand for this product family?

In addition, the general manager and sales and marketing may need to discuss whether the plans will yield the desired company objectives. Alternately, if the plans look too optimistic or pessimistic and the general manager considers them unrealistic, he may ask for more backup data supporting the plan or request a revision. The general manager must either accept the numbers or ask appropriate questions to ensure that the desired comfort level is achieved.

After the sales plan numbers have been reviewed and approved, the next step is to consider manufacturing and engineering's response to sales and marketing, so that consensus can be reached. The basic questions that must be answered in order to reach consensus are:

1. *Past performance: how have you done?* Where actual output has matched planned output, no discussion is necessary unless unusual steps were required, such as excessive overtime, or the department manager does not feel that the output rate can continue; for example, there may be some manpower or equipment levels that can't be maintained. What appears to be a satisfactory situation may therefore be one on the verge of causing a problem.

Where actual output has not matched planned output, the manufacturing and/or engineering managers should come forth with a reasonable explanation. More important, however, is the need to solve the problem. This becomes even more critical when the sales plan calls for an increase. If in the past the company has had to struggle to hit output rates, it's hard to have confidence that simply by trying harder, it can achieve a greater rate on a timely basis in the future.

2. *Future performance: can you maintain the current plan?* Certainly if the company has a history of underachievement, the manufacturing manager as well as the general manager should have a healthy skepticism about meeting the future commitment. The eternal optimist keeps expecting that problems will not repeat themselves. But unless there's a proven cure for the trouble, this confidence is built on quicksand. A better posture is to be a realist, one who can accurately assess his ability to perform and who has a positive attitude that improvements are definitely possible. Most important, there must be a program in place to prove it.

3. *Implementing changes to the plan—how soon and how costly?* Any change, of course, is possible, given time and/or money. Therefore, the best way to approach a requested change is to ask, "How soon and at what cost?" Short-term change means fewer choices, of which all are usually more costly than for long-term. Many companies attempt to meet their normal output commitments on a forty-hour workweek for two basic reasons: first, it's the most cost-effective, and second, it allows them to use overtime as a solution to quickly obtain more output when necessary.

With the use of rough-cut capacity planning, manufacturing and engineering can review the impact on their key resources of a requested change in output. This approach focuses attention where it's deserved, and often results in a quick response. There are times, however, when much further investigation is required, and it may not be possible to present a reliable answer by the time of the meeting. Depending on the urgency of the issue, it may simply be noted and carried to the following month's agenda, or it may require a special meeting prior to the regular meeting.

4. *Present alternatives—which is the best one?* The ideal situation would be for each of the support groups to be able to say, "Yes, we can satisfy the requested change without any problem." There will be times, however, when the ideal *doesn't* occur. In such cases the best response, from manufacturing, engineering, and purchasing's viewpoint, is apt to be: "Hold the course. We've overreacted in the past. Is it really that important anyway? Why don't we make the change in small steps at a slow pace?" That's not to say that members of these groups are simply foot-draggers, always skeptical of marketing's request. Rather, because they are measured by whether they fulfill their commitments and how economically they do so, they seek to present a plan that will achieve those objectives. They can also spell out the other alternatives, in each case estimating the consequences in terms of time and money. They may have some opinions on which of these plans make the most sense to them and why, but in the final analysis they recognize that their role is to do the same thing that marketing has done, namely, to step forward with recommendations.

5. New Products

Major new products are the hope of the future for the company. Too often, though, new products are treated as "second-class citizens," receiving insufficient planning and control. This results in surprises and delays as things fall between the cracks.

The following features of new products make them more tricky to deal with from a Sales & Operations Planning perspective:

1. It's more difficult to forecast demand—there's no history to utilize.
2. It's hard to determine the resources and standard hours until bills of materials and routings are available. You may not really know them until the product has been produced.
3. There are no timing precedents, so it may be difficult to deter-

mine what has to be done in order to hit the release dates.

4. A large cast of characters—many people and departments (design, engineering, manufacturing, production, purchasing, sales, finance, etc.)—is involved, so there is more to coordinate and more things can go wrong.

5. More engineering changes than mature products. This adds to instability.

6. When a new product replaces an old product, you often must go through a tough balancing act—deciding when to go with the new and stop the old. If the new product won't be ready as soon as originally planned and the old product still has a long lead time, you must plan accordingly, or you'll be late on the new *and* out of the old product. If that happens, you'll be doing your competitor a big favor.

To provide a proper focus on new products, many companies treat them as a separate family, even though they technically may be extensions or modifications of existing product lines. Other companies go even further and have a full-time master scheduler coordinating new products. There is no "right" way to categorize and manage new products, so you must choose the approach that seems most appropriate to your particular needs. Many companies improve their chances for on-time new product introduction by documenting and discussing the items that could be the biggest pitfalls.

We are not suggesting that all aspects of new product introduction get reviewed at the Sales & Operations Planning meeting; you'll also conduct regular new product review meetings to cover the nitty-gritty details. What should be brought to the Sales & Operations Planning meeting are the results of the regular new products meetings, stated in terms of the overall impact to the business.

6. Special Projects

It's important to discuss any special projects going on in the company that could have an impact on the general business, especially if multiple departments are involved. Such projects or events might affect other plans and should be discussed at the current Sales & Operations Planning meeting. Depending on the project, you may have to schedule a separate meeting to discuss the details, and summarize the impact at the next formal Sales & Operations Planning session. Events, dates, and projects to be discussed include:

- additional facilities
- implementing Just-in-Time/Total Quality Control
- changes in labor force
- new technology
- new distribution strategies
- new advertising programs
- new information systems
- reorganization
- acquisitions

7. Review of Decisions, Reading of Minutes

Just as the general manager kicked off the meeting by discussing any special issues, at this point it is important that he reiterate any issues that need emphasis. Once the detailed family reviews have been completed, it is important to make sure you tie together the general business issues with the day-to-day operations.

A review of what decisions were made can eliminate miscommunication and misunderstanding. The person who takes down the minutes should therefore summarize the decisions and read the minutes aloud before the meeting is adjourned. This way, everyone will know exactly what information will be distributed to the rest of the company, and disagreements about what was said can be ironed out before the material is published. We are amazed that these reviews are not common practice. Be assured, they work!

Many executives actually feel so strongly about the importance of circulating the results of the Sales & Operations Planning meeting that they require the general manager to sign them. At one company, the last official action of the Sales & Operations Planning meeting is the signing of the approved document by the president in a space provided on the cover sheet. The president doesn't sign until all critical issues have been dealt with to his satisfaction. Sometimes this means extra work following the meeting and postponement of the actual signing, but it does emphasize both the president's commitment and his charge to all members of the group that the issues critical to the success of Sales & Operations Planning must be considered before the document is accepted as operating policy.

8. Critique of Meeting

Even though this is number eight on the list, it may be the most important item; it appears last only because it is so in the sequence of the meeting. Companies just beginning Sales & Operations Planning must assess the quality of their meetings. The best way is to conduct a critique at the end of the session. The general manager should lead the critique, using a standard set of questions. These include:

- Was everyone prepared?
- Were the right people in attendance?
- Did we have the right level of detail?
- Were needed decisions made?
- Were we efficient with our time?
- Did we budget our time properly?
- What can we do to improve the process?

The final step is for the general manager to comment on whether he believes the process needs any improvement. After the Sales & Operations Planning process has been fine-tuned, it can still be worthwhile to do the critique on a quarterly basis—even the best-intentioned people can forget how to do things, unless they have a friendly reminder.

Tailor the Meeting Process to Your Own Company

The meeting agenda that we are proposing is intended to help you get started with the Sales & Operations Planning process. It is not meant to be constraining. Each company needs to customize the process to its own needs. Approached with the right attitude, this will come about quite naturally.

SUMMARY

- Give everyone the opportunity to describe any overriding issues at the start of the meeting. These issues should set the tone for the remainder of the meeting.
- Keep the performance reviews at an *overview* level; if the discussion gets bogged in detail too early, it will be hard for the team to grasp the big picture.
- Document and review assumptions in order to better understand what drives your business.
- The new product discussion should include a summary of potential pitfalls that could delay the launch or jeopardize its success.
- Meeting attendees should be encouraged to describe any special projects that affect the business. This part of the meeting should not be allowed to degenerate into a personal ''gripe'' session.
- Make sure everyone agrees with what will be published in the minutes that will be distributed throughout the company.
- When beginning Sales & Operations Planning, take a few minutes to critique the form and substance of the meeting. Even when the process is working smoothly, periodically evaluate meeting performance. This will help keep you on target and point out areas for improvement.

Chapter 7

Implementing the Process

To play great music, you must keep your eyes on a distant star.
(Yehudi Menuhin, American musician)

"There is no magic in the Sales & Operations Planning concept. It simply takes what common sense tells us is the correct way to interface all the various functions of a business, and provides us a regimented means with which to get the job done." That's how Robert Agan, president, and Thomas Connelly, vice president and assistant to the president of Hardinge Brothers, summarized the contribution Sales & Operations Planning makes to their company. Hardinge is a manufacturer of precision machines and attachments.

They cite three reasons for their success in putting Sales & Operations planning to work in their company:

> We feel the success of our Sales & Operations Planning program is the result of some very specific reasons. The first of these has to be commitment. Hardinge management decided at the very onset of this project that it would fully support the MRP II concept. As we became involved in the implementation of the program, it became quite obvious that the Sales & Operations Planning format affords top management a very effective way to demonstrate that commitment on a regular basis. Toward that end, we make it a practice always to thoroughly study the proposed plans prior to our meetings so that we are completely conversant with all of the issues likely to evolve during the meetings. When top man-

agement participates fully and actively in these meetings, there is no question on anybody's part that the promised commitment is in fact real. Once that is established, commitment at all levels of the company follows without fail.

Second, Hardinge uses Sales & Operations Planning effectively because we believe firmly in its concept. It provides the company with a specific format within which all of the various functions of the company can be united to achieve company goals. It provides a common ground where issues can be raised and responded to by all functions of the organization. It provides an excellent vehicle for management to get the issues and plans to all departments and to ensure all inputs are considered before decisions are made. It is, in fact, a very effective vehicle for intracompany communications.

Third, the Sales & Operations Planning concept is measurable. Effective use of it provides valuable input to overall business planning and forecasting techniques. At Hardinge we have directly linked the Sales & Operations Planning output to our budgeting and planning input to ensure our future course is determined using the best planning data available. The concept of measurability is inherent throughout Sales & Operations Planning. Forecasts are monitored for accuracy, production measured for performance to promise, product development measured for timeliness. And, for every measurement, a specific responsibility has been assigned to ensure performance. There is no thumb-passing with the Sales & Operations Planning concept. All members of the committee know this and, as a result, all tend to perform more conscientiously.

We're sold on the Sales & Operations Planning concept at Hardinge, and we're making it work for us. It's not easy. It takes time and commitment. It takes attention and education. But the benefits are obvious. To us, it is the Sales & Operations Planning concept that takes MRP II from simply being a production scheduling system to providing Hardinge with an effective way to run our business.

This is certainly an elegant testimonial of the value of Sales & Operations Planning at Hardinge Brothers. What Hardinge accomplished a few years ago is what we hope you can achieve after reading this chapter.

In the preceding chapters, we've covered the process of Sales & Operations Planning from concept through execution. Even though we have identified all of the ingredients of the process, we'd like to highlight the critical parts and describe how to convert them from ideas to reality. This discussion is aimed at the company that does not currently have a Sales & Operations Planning process in place, or the company that has a process resembling it, but wishes to improve on it.

IMPLEMENTING SALES & OPERATIONS PLANNING

The implementation of Sales & Operations Planning should be viewed as a project, and every good project should have a project manager. Who should that be? The general manager—after all, it's his process. While he should delegate data gathering and other clerical duties to his staff, he must assume leadership of the process right from the beginning to understand it thoroughly, to operate it effectively, and to accept responsibility for the results.

Here are the items that need to be considered to get Sales & Operations planning going within your company:

1. *Education.* Everyone who participates in the Sales & Operations Planning process must have sufficient understanding of it to know what's expected of them and how to make the maximum contribution. Sales & Operations Planning education can be accomplished through a combination of outside classes, video courses, and this text.[1] Generally, Sales & Operations Planning implementation is done in the context of MRP II implementation, so there are other basic topics that can be covered as well, such as the fundamentals of Manufacturing Resource Planning (MRP II), Just-in-Time, and Distribution Resource Planning (DRP).

2. *Company Policy.* After the initial education, the general manager and his staff should put together a policy statement describing the purpose of the Sales & Operations Planning activity. This should be a brief outline of what it is, why it is important, and what they expect to accomplish through it. Discussing these aspects as a group, and then reducing them to a few sentences, will be a productive exercise. It requires considerable work to boil down the overall concept and capture its essence. A sample of what should be included in a Sales &

[1] For more information on video education, refer to Appendix C.

Figure 7.1
Sales & Operations Planning Policy

Objective

The company is establishing a monthly Sales & Operations Planning process to support the business plan and to improve the ongoing operations of the company.

Process

The process will consist of a series of steps in which each department updates and communicates its plans to other departments in a prescribed sequence in preparation for a Sales & Operations Planning meeting, which will be held to review and approve all plans.

Schedules

The Sales & Operations Planning meeting will be held monthly and a schedule published for the next twelve months in order to ensure attendance.

Participants

The Sales & Operations Planning participants will include:
- general manager
- vice president of sales
- vice president of marketing
- vice president of operations
- vice president of engineering
- vice president of human resources
- vice president of finance
- director of materials

Other people will be called on to participate from time to time.

Families

The Sales & Operations Planning team will develop and agree on the most appropriate aggregate levels in order to communicate most effectively. The plans must support the customer (marketing) requirements and be convertible to financial and capacity requirements so that all departments can support the plans.

Horizon

The planning horizon must be long enough to ensure the required resources can be provided when needed and in an economical manner.

Figure 7.1 (*continued*)
Sales & Operations Planning Policy

Time Fences

The Sales & Operations Planning team will establish time fences to manage change.

Review

The Sales & Operations Planning team will constantly review and critique the Sales & Operations Planning process in order to keep it up to date with the needs of the business.

Signature

This policy is signed and agreed to by:
(all managers must sign)

Operations Planning policy is shown in Figure 7.1. The policy should be short and sweet, outlining what it is that should be done. It is not to be confused with how-to procedures. Also, note the signatures. They are the most important part of the document. The act of deciding to do Sales & Operations Planning and formally signing up for it begins the process as a team effort.

3. *Review of checklist.* Ollie Wight not only created Manufacturing Resource Planning, he also developed a checklist to enable a manager to evaluate how effectively MRP II is being used. The series of questions that made up the original checklist has gone through two revisions, reflecting the evolution that is occurring in the field of resource planning and scheduling. It's critical that a manager ask the right questions, so that he will know both what's happening in the present and where to direct his attention for the future. The answers to the questions can quickly bring problems to the surface so they can be corrected before developing into crises. Another equally important benefit for managers is that the checklist allows them to see which established goals have not been achieved.

The Oliver Wight group has recently identified what we feel are the "right questions" for the Sales & Operations Planning process. These are the questions we would ask when visiting a company in order to determine whether its Sales & Operations Planning process has been

Figure 7.2
Sales & Operations Planning

OVERVIEW QUESTION

1. *Sales & operations planning* is the management process that maintains the current operating plan in support of the business plan. The process consists of a formal meeting each month run by the general manager and covers a planning horizon adequate to plan resources effectively.

DETAIL QUESTIONS

1. There is a concise written sales & operations planning policy that covers the purpose, process and participants in the process.
2. Sales & operations planning is truly a process and not just a meeting. There is a sequence of steps that are laid out and followed.
3. The meeting dates are set well ahead to avoid schedule conflicts. In case of an emergency, the department manager is represented by someone who is empowered to speak for the department.
4. A formal agenda is circulated prior to the meeting.
5. The plans are reviewed by product family units of measure that communicate most effectively.
6. The new product development schedule is reviewed at the sales & operations planning meeting.
7. All participants attend the sales & operations planning meeting prepared. There are preliminary meetings by each department in preparation for the sales & operations planning meeting.
8. The presentation of information includes a review of both past performances and future plans for: sales, production, inventory, backlog, shipments and new product activity.
9. Inventory and/or delivery leadtime (backlog) strategies are reviewed each month as part of the process.
10. There is a process of reviewing and documenting assumptions about business and the marketplace. This is to enhance the understanding of the business and represents the basis for future projections.
11. Sales & operations planning is an action process. Conflicts are resolved and decisions are made and communicated.
12. Any changes—large and/or unanticipated, are communicated to other departments prior to the meeting.
13. Minutes of the meeting are ciculated immediately after the meeting.
14. The mechanism is in place to ensure that aggregate sales plans agree with detailed sales plans by item and by market segment or territory. There is a consensus from sales, marketing and operating management.

Figure 7.2 (*continued*)
Sales & Operations Planning

15. Time fences have been established as guidelines for managing changes. In the near-term, there is an effort to minimize the changes in order to gain the benefits of stability. In the mid-term range, changes up or down are expected but are reviewed to insure they can be executed. In long-term, less precision is expected but direction is established.

16. Tolerances are established to determine acceptable performance for: sales, engineering, finance and production. They are reviewed and updated. Accountability is clearly established.

17. The production plan is the driver of the master schedule and is supported by a procedure which defines a summarization to insure that they are in agreement.

18. There is an ongoing critique of the process.

designed correctly and is being operated properly. We recommend that you use this list in designing your process and periodically auditing your operations. (See Figure 7.2).

Some companies may want to add questions to the checklist so that it probes even deeper into their Sales & Operations Planning process. Similarly, some companies may wish to drop questions that don't apply to their operation. If you choose to drop questions, do so cautiously; they were derived from the many years of experience of the members of our group, and we feel that they represent important issues that apply to *all* manufacturing companies. Before excluding any of the questions from the checklist, the general manager should be absolutely certain that these questions are not appropriate to his company.

As part of the preparation for this book, we conducted a survey among companies that have excellent Manufacturing Resource Planning systems. Twenty questions were directed at the Sales & Operations Planning process. To each question, we asked the participants to provide two answers: whether they used the function and whether they felt the function is important. Eighty-six companies responded to the survey; an analysis of their answers appears in Appendix A.

The survey contains many of the same questions included in the checklist in Figure 7.2. We think you'll find the survey results informative, because they indicate what a number of companies are actually doing as well as what they should be doing. We also recommend that the implementation team—the general manager and his staff—review both the checklist and the survey. They can serve as a basis for discussion, helping you achieve the dual objectives of developing an effective Sales & Operations Planning process and putting in place an ongoing measurement system.

4. *Brainstorming Meetings.* Following the education and policy steps, one or possibly more meetings should be held to discuss the design of the Sales & Operations Planning process. During these sessions, the participants should explore the following issues: definition of families, unit of measure, and planning horizon. Almost always, there will be disagreements regarding each of these subjects; rarely do we see consensus at this stage. On the one hand, a thorough airing of the differences should be encouraged, to ensure that all aspects have been understood and included, but at some point an initial agreement should be reached. This permits the group to go on to the next stage, using the process and judging the design by hindsight.

As with all other parts of Sales & Operations Planning, there will be ongoing modifications. The initial design phase should be considered in this light. Even if there is initial consensus, subsequent experience will bring about a number of improvements.

Topics that should be included in the brainstorming and design phase include:

• Discussion of the format and the information on the reports. Do you have the right information to make an informed decision? Is the level of details on the reports appropriate? Is there agreement as to where the data will come from?
• Discussion of systems and data processing resources. What is required of systems and data processing to produce the desired reports, and how quickly can they accomplish the tasks?
• Exploration of whether an interim spreadsheet should be put into place, and whether it would save significant time.
• Documentation of assumptions and vulnerabilities. Start writing down your assumptions about the marketplace, the economy, the internal environment within your company, and other factors that can influence the decision-making process.
• Identification of what should be included in the Sales & Operations Planning process kit. Each department should list its responsibilities

for both the premeeting phase and the formal meeting. The group should then decide what the kit will contain.

5. *Determining where you want to meet your customer.* Two approaches are available here. The first is to accept your current practice as being appropriate, and therefore incorporate it in your Sales & Operations Planning process. This means that you should identify which families should be treated as make-to-stock, which families would be considered make-to-order, and what additional families exist today, such as service parts, emergency orders, etc., as we described in earlier chapters. The other approach is to use the implementation phase as an opportunity to analyze current practice in terms of your company's strategies. Are they what they should be? Would a change offer any competitive advantage? For each family, there will be a need to establish targets in terms of finished-goods inventory levels and backlog of customer orders.

6. *Time fences and rough-cut capacity planning.* It is unlikely that it will be immediately obvious where time fences should be established for managing changes or which work centers are critical to rough-cut capacity planning. Do not let this consume an unreasonable amount of review or cause undue apprehension about starting. Rather, strive for a workable decision at this time, while recognizing that these issues will need further attention. As part of the ongoing Sales & Operations Planning process, it would be appropriate to assign detailed studies of the alternatives and to hold further discussions.

7. *Scheduling meetings.* This includes determining dates for all meetings, including the pre–Sales & Operations Planning meetings for each department. The cutoff times and steps for gathering the required data must be established. The format in which to present the data must be developed.

KEEPING THE PROCESS GOING

The initial meetings are apt to uncover two problem areas: (1) the data's accuracy, timing, and format; and (2) performance measurements—the tendency of people to be overly defensive when performance in their area is less than expected. Although it is normal to experience difficulties in both of these areas, the general manager must continually reset the course and work at providing the right atmosphere to ensure that the short-term problems can be overcome and that the planning team can get on with the important task of running the business.

The best way to refine your Sales & Operations Planning process is to honestly and regularly critique it. Care should be taken to make sure that everyone understands that the critique is an opportunity to learn how to make the process work more smoothly and efficiently, rather than a means for people to point out that someone did something wrong. Whenever a forum is operated in a positive manner, continuous improvements are likely.

It is not an easy task to conduct efficient meetings. The general manager who feels he's currently doing such a good job that there's no need to periodically assess his performance and seek the advice of others is handicapping himself and therefore his company. In the book *Human Side of Enterprise*, Douglas McGregor lists the characteristics that distinguish productive meetings from poor ones. These are so critical for the success of Sales & Operations Planning that we've included them in Figure 7.3.

Figure 7.3

---

1. The "atmosphere," which can be sensed in a few minutes of observation, tends to be informal, comfortable, relaxed. There are no obvious tensions. It is a working atmosphere in which people are involved and interested. There are no signs of boredom.
2. There is a lot of discussion in which virtually everyone participates, but it remains pertinent to the task of the group. If the discussion gets off the subject, someone will bring it back in short order.
3. The task or the objective of the group is well understood and accepted by the members. There will have been free discussion of the objective at some point until it was formulated in such a way that the members of the group could commit themselves to it.
4. The members listen to each other! The discussion does not have the quality of jumping from one idea to another unrelated one. Every idea is given a hearing. People do not appear to be afraid of being foolish by putting forth a creative thought even if it seems fairly extreme.

Figure 7.3 (*continued*)

5. There is disagreement. The group is comfortable with this and shows no signs of having to avoid conflict or to keep everything on a plane of sweetness and light. Disagreements are not suppressed or overridden by premature group action. The reasons are carefully examined, and the group seeks to resolve them rather than to dominate the dissenter.

   On the other hand, there is no "tyranny of the minority." Individuals who disagree do not appear to be trying to dominate the group or to express hostility. Their disagreement is an expression of a genuine difference of opinion, and they expect a hearing in order that a solution may be found.

   Sometimes there are basic disagreements which cannot be resolved. The group finds it possible to live with them, accepting them but not permitting them to block its efforts. Under some conditions, action will be deferred to permit further study of an issue between the members. On other occasions, where the disagreement cannot be resolved and action is necessary, it will be taken but with open caution and recognition that the action may be subject to later reconsideration.

6. Most decisions are reached by a kind of consensus in which it is clear that everybody is in general agreement and willing to go along. However, there is little tendency for individuals who oppose the action to keep their opposition private and thus let an apparent consensus mask real disagreement. Formal voting is at a minimum; the group does not accept a simple majority as a proper basis for action.

7. Criticism is frequent, frank, and relatively comfortable. There is little evidence of personal attack, either openly or in a hidden fashion. The criticism has a constructive flavor in that it is oriented toward removing an obstacle that faces the group and prevents it from getting the job done.

8. People are free in expressing their feelings as well as their ideas both on the problem and on the group's operation. There is little pussyfooting, there are few "hidden agendas." Everybody appears to know quite well how everybody else feels about any matter under discussion.

9. When action is taken, clear assignments are made and accepted.

10. The chairman of the group does not dominate it, nor on the contrary, does the group defer unduly to him. In fact, as one observes the activity, it is clear that the leadership shifts from time to time, depending on the circumstances. Different members, because of their knowledge or experience, are in a position at various times to act as "resources" for the group. The members utilize them in this fashion and they occupy leadership roles while they are thus being used.

   There is little evidence of a struggle for power as the group operates. The issue is not who controls but how to get the job done.

Figure 7.3 (*continued*)

11. The group is self-conscious about its own operations. Frequently, it will stop to examine how well it is doing or what may be interfering with its operation. The problem may be a matter of procedure, or it may be an individual whose behavior is interfering with the accomplishment of the group's objectives. Whatever it is, it gets open discussion until a solution is found.

SUMMARY

In our experience, by following the above implementation steps a company will be able to make effective use of the Sales & Operations Planning process within three to six months. If you find this is not occurring at your company, a major problem is likely present. Instead of continuing with little, if any, progress and becoming discouraged, the general manager should postpone the meetings and review the situation. The diagnosis should separate information problems from people problems. Even if both categories exist, it's easier to deal with them separately. Since the concept behind Sales & Operations Planning is largely straightforward, logical, and based on common sense, the managers will always be able to uncover the causes and solve the problems.

You will know Sales & Operations Planning is working well as soon as someone says, "We should have always been doing it this way." Whenever we have an opportunity to sit in on a Sales & Operations Planning meeting, we can quickly judge its effectiveness; it is always very apparent, as well as impressive, when a group of individuals works as a confident team. Three ingredients generate such meetings: timely and accurate data, the right attitude, and a correct understanding of the process. Sales & Operations Planning always works when this combination comes together.

Appendix A
Sales & Operations Planning Survey

An effective Sales & Operations Planning process is an integral part of running a manufacturing company at a Class A level. The term "Class A" was coined by the late Ollie Wight, who created a checklist for executives to evaluate how well they were using Manufacturing Resource Planning (MRP II). Initially, it consisted of twenty questions, and subsequently was revised to reflect the evolution of the process, which continues to take place today. By answering the questions on the checklist, an executive can rate his operation—his MRP II system—to determine how well he's using it.

"A" means very well; "B" is good; "C" is average; and "D" is poor. Of the total number of companies using MRP II, a select number excel at operating MRP II, utilizing it to its full potential, while the majority would be considered Class B or C users. Companies below the Class A level are certainly not failures; in fact, the B users are getting outstanding results. Many C users have also achieved benefits that have exceeded their investment.

Having an accepted standard for measuring performance is helpful. You can judge where you are and you can use the established benchmarks to motivate your group to perform better. Becoming Class A does not mean there is nothing left to accomplish. It simply reflects a

125

Figure A-1

Number Of Employees

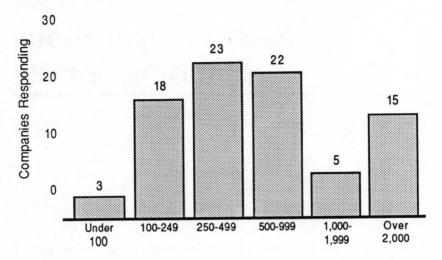

high degree of achievement. Executives of Class A companies are likely to be the most determined to continue to improve in the area of key operating measurements: better customer service, improved quality, reduced costs, and increased inventory turns.

To survey how companies were using Sales & Operations Planning, we turned to the "pros"—our list of Class A and B users. We wanted to know not only what they were doing, but how important they considered it. To our survey of twenty questions, sixty Class A and twenty-six Class B MRP II companies responded. These eighty-six companies, measured by both annual sales and employees, represented a range of sizes, as shown in Figures A.1 and A.2.

Reviewing the results of the survey can be an educational experience, as it reflects what other people are doing with Sales & Operations Planning. Drawing proper conclusions is not as straightforward, however. The survey participants may not be doing the right thing, or it

Figure A-2
Annual Sales

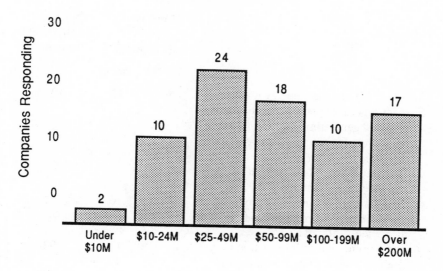

could be right for them but not for you. The particular value of this survey lies with the outstanding credentials of the people we sent it to. They are performers, having done what many of the readers of this book hope to do, and they are doing it extremely well. To us, these companies are excellent role models and we feel it should be a good one for you as well.

Each of the twenty questions represented a Sales & Operations Planning activity. We wanted to know the following: are you doing this activity, and do you consider it important? When we analyzed the answers from the eighty-six companies, we found there were three groupings in terms of popularity. Eight questions received a very high rate of "yes" answers, ranging from 84 percent to 100 percent. There were six questions in the middle range, to which 64 to 78 percent responded positively. The third category contains the remaining six questions, and the range of responses was from a low of 48 percent to a high of 56 percent.

Here are the activities that were the most popular:

	DOING IT TODAY (percentage)			IMPORTANT ENOUGH TO BE MEASURED (percentage)
	Yes	No	Partial	
1. S&OP is treated as the key management process supporting the business plan and generating a rolling "current" operating plan for the fiscal year and beyond.	88	0	12	100
2. The S&OP consists of at least one formal meeting per month.	90	5	5	88
3. The plans are reviewed by product family in the units of measure that communicate most effectively.	87	3	10	81
4. The presentation of information includes a review of both past performance and future plans for sales, production, inventory, backlog, and shipments.	84	3	13	93
5. The planning horizon is sufficient for business planning, capacity planning, vendor scheduling, and performing "what-if" analysis.	90	1	9	87
6. Time fences have been established as guidelines for managing changes. In the near term, there is an effort to minimize changes in order to gain the benefits of stability. In the midterm range, changes up or down are expected but are reviewed to ensure they can be executed. In the long term, little precision is expected; simply the need to establish direction is discussed.	84	3	13	89
7. The production plan is supported by a master scheduling procedure, which includes a summarization to ensure that they are in agreement.	94	0	6	95

8. S&OP is an action process. Conflicts are resolved and decisions are made and communicated.	88	0	12	86

There are no surprises with these answers. The vast majority of the eighty-six companies are actively performing these Sales & Operations Planning functions, and rate them extremely high in terms of importance. If anything, it would be a surprise to discover that these eight questions any less.

The middle category of popularity included the following activities:

	DOING IT TODAY (percentage)			IMPORTANT ENOUGH TO BE MEASURED (percentage)
	Yes	No	Partial	
9. S&OP is truly a process and not just a meeting. There is a sequence of steps that are laid out and followed.	77	3	20	79
10. The general manager and department managers attend the S&OP meeting. The meeting dates are set well in advance to avoid schedule conflicts. In case of an emergency, the department manager is represented by someone who is empowered to speak for the department. The general manager runs the meeting.	70	6	24	83
11. Any major and/or unanticipated changes are communicated to other departments prior to the S&OP meeting.	67	8	25	70
12. A rough-cut capacity planning mechanism is in place to predict the consequences of various plans on the key resources.	72	13	15	82
13. Inventory and/or delivery lead time (backlog) strategies are reviewed each month as part of the process.	78	9	13	83

14. The mechanism is in place to ensure   64 13 23   83
    that aggregate sales plans agree with
    detailed sales plans by item and by
    market segment or territory. There is a
    consensus from both sales and
    marketing.

For three of the above activities, a significant gap exists between the percentage of companies that regularly perform the activity and the percentage that consider it important. Number 10 shows that while the general manager attends S&OP meetings for 70 percent of the companies, 83 percent felt it was important he do so. Our experience supports this—companies where the general manager does not actively participate envy those companies where he does. Number 12 indicates that while it would be helpful to have a rough-cut capacity planning tool, it is not today available in a number of companies. Number 14 points out a potentially serious problem. There is great danger for companies that do not reconcile aggregate plans with detail plans, because they may be out of step with each other. An executive who wanted tight controls would not rely on faith, hoping that the two would not deviate greatly. Rather, he would insist that they be in lockstep, and to ensure that they were, he would have appropriate measurements.

The fewest companies reported performing the following activities:

	DOING IT TODAY (percentage)			IMPORTANT ENOUGH TO BE MEASURED (percentage)
	Yes	No	Partial	
15. There is a concise, written S&OP policy. This covers purpose, process, and participants.	56	14	30	76
16. All participants must attend the S&OP meeting prepared. There are preliminary meetings of each department in preparation for the S&OP meeting.	56	13	27	73
17. A formal agenda is circulated prior to the meeting.	55	41	4	55

18. Tolerances are established to determine acceptable performance for sales, engineering, finance, and production. They are reviewed and updated. Accountability is clearly established.   48   16   36   87

19. There is a process of reviewing and documenting assumptions about business and the marketplace. This is to enhance the understanding of the business and represents the basis for future projections.   56   19   28   57

20. Minutes of the meeting are circulated immediately after the meeting.   55   31   14   65

Number 15 shows the second-largest difference overall—a gap of 20 percent—between the percentage of companies regularly practicing the activity (56 percent) and the percentage that feel it is important (76 percent). We advocate viewing this discrepancy in the light of the age-old advice "Do as I say, not as I do." There are many companies that operate smoothly without written procedures to guide them. If you are starting up Sales & Operations Planning, we would encourage you not only to think it through, but to write it out to help establish an effective process. Use hindsight to determine whether such things as procedures, agenda, and minutes are valuable or not.

Regarding Number 16, we find that smaller companies have less of a need for preliminary meetings. In these companies, everyone wears several hats, and has more knowledge of what's occurring in other areas as well as detailed knowledge of their own area. The communications are therefore different. In a larger company, the executives are more removed from the details, and the communication is far more challenging as the company increases in size. In such cases, the preliminary meetings are a means of addressing this challenge productively.

Numbers 17 and 19 possess the distinction of having received few positive responses with regard to either practice or importance. Number 17, like Number 20, is in a category that could be considered "administrative," and the responses to both indicate that a great many companies walk into their meetings without a formal agenda and leave without minutes having been taken or circulated. One could make a case that this is as it should be, that actions speak louder than words. Our concern for companies that want to follow this approach centers on the use of people's time in the meetings, as well as the effectiveness of their action following the meetings. The agenda serves as an organizer to ensure that the group is addressing the important issues. If a company finds there are hardly any deviations from one meeting to another, that one agenda looks the same as all others, then it would have a legitimate case for either dusting off the master copy each time or discontinuing the agenda altogether. Companies in which changes are reflected on the agenda will find that it serves a helpful purpose.

Number 18 shows the largest gap, 39 percent, between those who perform the activity and those who consider it important. Note, however, that a significant number of companies partially practice this activity. The absence of tolerances generally leads to a debate over where "good" ends and "bad" begins. Much like a design engineer, you should present the desired dimension with an upper and lower limit, reflecting acceptable tolerances. Likewise, ranges should be established for the performances of each functional area attending the Sales & Operations Planning meeting. This quickly establishes whether you are in or out of control. When an area's performance is inside tolerance, little time should be spent on it, whereas those categories that are out of control need attention. It is important to determine both the causes and the corrections for this unsatisfactory situation.

We were not surprised by the answers to Number 19. Our experience in working with a great many manufacturing companies confirms that the whole area of documenting assumptions is poorly understood. What they are, why they are important, and how to establish them are not widely recognized. Although we have no way to back up a prediction, it is our feeling that if those companies in the survey not utilizing assumptions were to become more familiar with them, a much higher percent would use them.

Number 20 is, as we noted earlier, one of the administrative-type issues that received a less than enthusiastic response. Our main worry with companies not recording decisions concerns people's memories. Will everyone retain a correct and common image of what was decided? Will they remember this during the weeks prior to the next meeting?

Too often, we've seen short-lived recall. It resembles the child's game where you whisper a few words in one person's ear and ask them to pass it on. About four people later, out comes something that doesn't resemble what was originally stated. Although it is not intentional, one of our characteristics is to better remember what we want to hear than the what we don't, or to remember a piece of a conversation rather than the whole. If you find there is no disagreement among the group members as to what was decided, then there is less need for writing it down and circulating it. Better to put in the extra effort initially, and let it serve as an insurance policy to help with both the communication and memory issues, than to risk the consequences of everyone being slightly out of step.

A seasoned veteran with years of experience does a better job of relating what he is doing today than of remembering how he got there. Thus, caution is required for those of you beginning this process and learning from people who have been doing it for a long period of time. All processes evolve with time. When beginning, resist the shortcuts. The extra effort will not hurt the process, and will likely improve its effectiveness.

The Fundamentals of Manufacturing Resource Planning and Just-in-Time/ Total Quality Control

Edited by Dana W. Scannell
Vice President and Publisher
Oliver Wight Limited Publications, Inc.

To be truly competitive, manufacturing companies must deliver products on time, quickly, and economically. Manufacturing Resource Planning (MRP II) and Just-in-Time/Total Quality Control (JIT/TQC) have proven to be essential tools in achieving these objectives.

Their capabilities offer a means for effectively managing the required resources: materials, labor, equipment, tooling, engineering specifications, space, and money. For each of these resources, Manufacturing Resource Planning can calculate what's required, when it's needed, and how much is needed. Having matched sets of resources at the right time and the right place is essential for an economical, rapid response to customer demands. Just-in-Time/Total Quality Control forces the elimination of waste and removal of obstacles that get in the way. MRP II's strength is in planning and control; JIT/TQC's strength is in execution and continuous improvement. Both are essential to be a winner in manufacturing.

PART ONE—MANUFACTURING RESOURCE PLANNING (MRP II)

The logic of MRP II is quite simple; it's in every cookbook. The sales & operations plan says that we're having Thanksgiving dinner on the third Thursday in November. The master schedule is the menu, including turkey, stuffing, potatoes, squash, vegetables, and all the trimmings. The bill of material says, "Turkey stuffing takes one egg, seasoning, bread crumbs, etc." The routing says, "Put the egg and the seasoning in a mixer." The mixer is the work center where the processing is done.

In manufacturing, however, there's a lot more volume and a lot more change. There isn't just one product, there are many. The lead times aren't as short as a quick trip to the corner store, and the work centers are busy. Thanksgiving won't get rescheduled, but customers sometimes change their minds. The world of manufacturing is a world of constant change, and that's where MRP II comes in.

The elements that make up an MRP II operating and financial planning system are shown in figure A-1. We'll briefly walk through each to get an understanding of how MRP II operates.

BUSINESS PLANNING

Business planning represents the overall plan for the company, taking into account the needs of the marketplace (customer orders and forecasts), the capabilities within the company (people skills, available resources, technology), financial targets (profit, cash flow, and growth), and strategic goals (levels of customer service, quality improvements, cost reductions, productivity improvements, etc.). The business plan is expressed in dollars and lays out the long-term direction for the company. The general manager and his staff are responsible for maintaining the business plan.

SALES & OPERATIONS PLANNING

Sales & operations planning addresses that part of the business plan which deals with sales, production, inventories, and backlog. It's the operational plan designed to execute the business plan. As such, it is

Figure A-1
Manufacturing Resource Planning (MRP II)

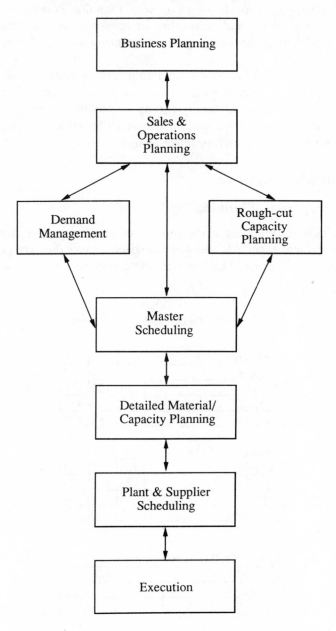

stated in units of measure such as pieces, standard hours, etc., rather than dollars. It's done by the same group of people responsible for business planning in much the same way. Planning is done in broad categories of products and establishes an aggregate plan of attack for sales and marketing, engineering, manufacturing and purchasing, and finance.

A simple example of a sales & operations plan is contained in figure A-2. In the sales & operations plan, the sales plan and production plan are aggregated by product family (group of products, items, options, features, etc.), showing the ending inventory/backlog (depending on whether it's a make-to-stock or make-to-order product).

DEMAND MANAGEMENT

Forecasting/Sales Planning

Forecasting/sales planning is the process of predicting what items the sales department expects to sell and the specific tasks they are going to take to hit the forecast. The sales planning process should result in a

Figure A-2

Sales Plan	-3	-2	-1
Planned	120	120	120
Actual	124	123	138
Diff.	+4	+3	+18
Cum. Diff.		+7	+25

Production Plan

Planned	125	125	125
Actual	121	119	118
Diff.	-4	-6	-7
Cum. Diff.		-10	-17

Inventory/Backlog

Planned		103	108	113
Actual	98			
Diff.				

monthly rate of sales for a product family (usually expressed in units identical to the production plan), stated in units and dollars. It represents sales and marketing's commitment to take all reasonable steps to make sure the forecast accurately represents the actual customer orders to be received.

Customer order entry and promising

Customer order entry and promising is the process of taking incoming orders and determining specific product availability and configuration. It results in the entry of a customer order to be built/produced/shipped, and should also tie to the forecasting system to net against the projections. This is an important part of an MRP II system; to look at the orders already in the system, review the inventory/backlog, available capacity, and lead times, and then determine when the customer order can be promised. This promise date is then entered as a customer commitment.

ROUGH-CUT CAPACITY PLANNING

Rough-cut capacity planning is top management's rough estimate of determining what it will take to achieve the plans discussed in the sales & operations planning meetings and if it is realistic. The process relies on aggregate information in hours and units to highlight potential problems in the plant, engineering, finance, etc. prior to the proposed schedule being approved. It takes the machine and labor hours available and balances that with what the company is trying to achieve in production.

MASTER SCHEDULING

Master scheduling is a detailed statement of what products the company will build, stated by individual item rather than families. It is broken out into two parts—how many and when. It takes into account existing customer orders, forecasts of anticipated orders, current inventories, and available capacities. This plan must extend far enough into the future to cover the sum of the lead times to acquire the necessary resources. The master schedule must be laid out in time periods of weeks or smaller in order to generate detailed priority plans for the execution departments to follow. The sum of what's specified in the master schedule must reconcile with the sales & operations plan for the same time periods. Figure A-3 is an example of a master schedule including the available-to-promise calculation used in customer order promising.

Figure A-3

Master Schedule

Week		1	2	3	4
Forecast		20	20	20	20
Customer Orders		15			
Master Schedule		40		40	
Inventory	0	20	0	20	0
Available to Promise		25		40	

DETAILED MATERIAL/CAPACITY PLANNING

Material Requirements Planning (MRP)

Material requirements planning starts by determining what components are required to execute the master schedule, plus any needs for service parts/spare parts. To accomplish this, MRP requires a bill of material to describe the components that make up the items in the master schedule and inventory data to know what's on hand and/or on order. By reviewing this information, it calculates what existing orders need to be moved either earlier or later, and what new material must be ordered. Figure A-4 is an example of an MRP display for an item.

Figure A-4

Week		1	2	3	4
Requirements		80		80	
On Order					200
Inventory	100	20	20	-60	140
Planned Orders					

Capacity Requirements Planning (CRP)

Capacity requirements planning takes the recommended needs for manufactured items from MRP and converts them to a prediction of how much capacity will be needed and when. A routing which defines the operations involved is required, plus the estimate of time required for each. A summary by key work center by time period is then presented to compare capacity needed to capacity available. Figure A-5 is an example of a capacity plan for a work center.

Figure A-5

Week	1	2	3	4
Capacity (Required)	90	90	90	110
Capacity (Available)	90	90	90	90
Overload/Underload	0	0	0	-20

PLANT AND SUPPLIER SCHEDULING

Plant scheduling is two-way communication with the factory. Utilizing MRP, the plant is advised of which jobs have been scheduled, where they are located, and the priority of each. Furthermore, a company must also monitor the flow of capacity by comparing how much work was to be completed versus how much has actually been completed. This technique is called input-output control, and its objective is to ensure that what is actually occurring matches with what should be occurring in terms of capacity.

There are two basic ways for companies to arrange their production facilities: job shop and flow shop. A brief description of each follows.

Job Shop

A job shop is where the resources are grouped by like type. The classic example of this approach is a machine shop, where the lathes are in one area, the drills in another, etc. Work moves from work center to work center based on routings unique to the individual items being produced. Job shop scheduling is called dispatching, and there is a daily dispatch list for each work center. The dispatch list is simply a schedule of the work to be done in the work center, sorted in due date sequence. An example of a dispatch list is shown in figure A-6.

Figure A-6

Shop Order	Part Number	Operation Sequence Number	Qty.	Operation Start	Operation Due	Order Due	Std. Hrs. Set Up	Std. Hrs. Run	Status
17621	91762	020	50	8/13	8/16	8/25	—	3.5	R
18430	98340	030	500	8/13	8/18	8/31	4.3	14.3	H
18707	78212	010	1100	8/16	8/18	9/6	1.1	18.2	T
18447	80021	020	300	8/17	8/19	8/28	1.5	9.0	T
19712	44318	020	120	8/24	8/26	9/10	3.3	15.1	

Flow Shop

A flow shop is where the resources are grouped by their sequence in the process. An example of this type of environment is an automobile assembly line, oil refineries, a filling line in a consumer package goods plant, or a manufacturing cell. Work moves through the process automatically, so there is no need to use formal routings and move from operation to operation. Basically the raw material(s) are at one end, and the finished product comes out the other. To schedule the shop in this type of environment, a line/cell schedule is used. The line/cell list shows the jobs to be started in the priority sequence as they should begin, taking into account various run times and expected outcomes. An example of a line/cell list is shown in figure A-7.

Figure A-7

	WEEK 1 M	WEEK 1 T	WEEK 1 W	WEEK 1 TH	WEEK 1 F	WEEK 2 M	WEEK 2 T	WEEK 2 W	WEEK 2 TH	WEEK 2 F	WEEK 3	WEEK 4
BLEND X:												
REQMTS.						1100	1100	1100	1100	1100	4400	
SCHED.					1100	1100	1100	1100	1100	1100	3300	
BLEND Z:												
REQMTS.	1050	1050	1050	1050	1050							5250
SCHED.	1050	1050	1050	1050							1050	4200

Supplier Scheduling

Suppliers also need valid schedules. Supplier scheduling replaces the typical and cumbersome cycle of purchase requisitions and hard copy purchase orders. Within MRP II, the output of MRP for purchased items is summarized and communicated directly to suppliers. Long term contracts define prices, terms, conditions, and total quantities, and supplier schedules authorizing delivery are generated and communicated at least once per week, perhaps even more frequently in certain Just-in-Time/Total Quality Control environments. Supplier scheduling includes those changes required for existing commitments with suppliers—materials needed earlier than originally planned as well as later—plus any new commitments that are authorized. To help suppliers do a better job of long-range planning so they can better meet the needs of the company, the supplier scheduling horizon should extend well beyond the established lead time. An example of a supplier schedule is shown in figure A-8.

Figure A-8

RAW MATL R:	WEEK 1					WEEK 2					WEEK 3	WEEK 4
	M	T	W	TH	F	M	T	W	TH	F		
REQMTS.	1050	1050	1050	1050	1100	1100	1100	1100	1100	1100	4350	4200
INV. 3600 LBS.	2550	1500	450	9400	8300	7200	6100	5000	3900	2800	8450	4250
SCHED.			10,000								10,000	

EXECUTION AND FEEDBACK

The execution phase is the culmination of all the planning steps, where the items or processes are actually created. Any problems with materials or capacity are resolved through interaction between the plant and the planning department. This is done on an exception basis, and feedback will only be necessary when some part of the plan cannot be executed. This feedback consists of stating the cause of the problem and the best possible new completion date. This information must then be analyzed by the planning department to determine the consequences. If an alternative cannot be found, the planning department should feed the problem back to the master scheduler. Only if all other practical choices have been exhausted should the master schedule be altered. If the master schedule is changed, the master scheduler owes feedback to sales if a promise date will be missed, and sales owes a call to the customer if an acknowledged delivery date will be missed.

By integrating all of these planning and execution elements, MRP II becomes a process for effectively linking long-range aggregate plans to short-term detailed plans. From top to bottom, from the general manager and his staff to the front-line operators, it ensures that all activities are in lockstep to gain the full potential of a company's capabilities. The reverse process is equally important. Feedback goes from bottom to top on an exception basis—conveying unavoidable problems in order to maintain valid plans. It's a rack-and-pinion relationship between the top-level plans and the actual work done in the plant.

In addition to MRP II's impact on the operations side of the business, it has an equally important impact on financial planning. By including the selling price and cost data, MRP II can convert each of the plans into dollars. The results are time-phased projections of dollar shipments, dollar inventory levels, cash flow, and profits.

Incorporating financial planning directly with operating planning produces one set of books. The same data is driving both systems—the only difference being the unit of measure. Too often financial people have had to develop a separate set of books as they couldn't trust the operating data. Not only does this represent extra effort, but much judgment and guesswork have to be applied to determine the financial projections.

In addition to operating and financial planning information, simulations represent the third major capability of MRP II. The ability to produce information to help answer "what if" questions and to contribute to contingency planning is a valuable asset for any manager to have. What if business goes up faster than expected; what if it goes up as planned, but the mix of products shifts; what if our costs increase, but our prices do not; do we have enough capacity to support our new products and maintain sales for current ones? These are common and critical issues that constantly arise in all manufacturing companies. The key part of the management job is to constantly think through alternative plans. With MRP II, people can access the data needed to help analyze the situation, play "what if," and, if required, initiate a better plan.

PART TWO—JUST-IN-TIME/TOTAL QUALITY CONTROL

The Just-in-Time/Total Quality Control process seeks to eliminate waste in all areas. Waste is defined as any non value-added activity, and further classified into two categories, necessary and unnecessary. An example of necessary waste is double-entry bookkeeping, which doesn't add value to the product but is required. Unnecessary waste would be excess steps or paperwork that may be in the accounting process. Essentially, any resource that is not actively involved in a process that adds value is in a waste state. Therefore, waste includes all inventory not actually being worked on, all inventory not needed at a particular moment in time, all inspection to detect defective products, and all movement from one operation to another. Waste exists in all aspects of the business, and should be eliminated wherever possible because it not only adds cost without value, but also because it constrains the ability to respond economically to change.

The ability to respond to change can be measured by the speed at which we can economically convert materials into shipments. As a result, Just-in-Time/Total Quality Control drives us toward high velocity manufacturing. It often begins in manufacturing, but quickly encompasses all organizations in the company, as well as suppliers and customers.

Just-in-Time/Total Quality Control also means repeated reductions in order quantities, safety stocks, queues, rejects, set-ups, transactions, complexity of the product, number of suppliers, time in order entry, days of customer lead time, warranty claims, and customer returns, all of which occur over and over, day after day.

We cannot, however, suddenly eliminate waste, or economically have zero inventories or lot sizes of one overnight. We cannot immediately eliminate inspection without adverse consequences. In fact, blindly following the process can lead to higher costs, worse quality, and unsatisfactory product deliveries.

With that discussion of what Just-in-Time/Total Quality Control is, let's talk about how to correctly implement it.

THE JUST-IN-TIME/TOTAL QUALITY CONTROL PROCESS

The process of Just-in-Time/Total Quality Control is to eliminate waste, but we must first understand it before we can use it to economically improve quality, delivery, and cost. As shown in figure A-9, the Just-in-Time journey begins and progresses by learning how to economically manufacture "one less at a time." The one-less-at-a-time process is designed to:

• Continuously expose and prioritize wasteful constraints.

• Stimulate everyone to think effectively about solutions to the prioritized constraints.

• Provide visual feedback on our progress.

The one-less process is as critical to Just-in-Time/Total Quality Control as the computer is to Manufacturing Resource Planning. We can eliminate waste without the one-less process, but we are missing the continuous improvement driver and we are not likely to get as far nor travel as fast. Most companies try to eliminate waste. "One less at a time" is what makes it different from just operating the same old way.

Just-in-Time/Total Quality Control contains a technique called *kanban* which sets an upper limit on inventory. Kanban is a Japanese word that means "card," "signal," or "visual record." Each kanban in the system authorizes pieces of inventory, or orders in order entry, or specifications in design engineering, etc. The total number of kanbans in the system limits the total amount of inventory in the manufacturing pipeline. Simply, the essence of the kanban system is like the old-fashioned milk delivery system. Whenever the milkman would see an empty bottle of milk, he would replace it with a full one. Unfortunately, to some people, kanban cards may imply a new scheduling system. While kanban controls will replace or simplify conventional shop floor work order systems, the main benefit by far of the Just-in-Time/Total Quality Control process occurs when we continue to make shipments, while slowly removing kanbans (wasteful inventory), to expose constraints to higher manufacturing velocities. Just-in-Time/Total Quality Control is designed to expose constraints, but wishful thinking will not make the constraints disappear; we must use adequate problem-solving tools.

Just-in-Time/Total Quality Control involves all areas of the company. However, different companies with varying needs will choose different parts of the process, depending on where they are feeling the most pain.

Figure A-9   The Just-in-Time Process: "One Less at a Time"*

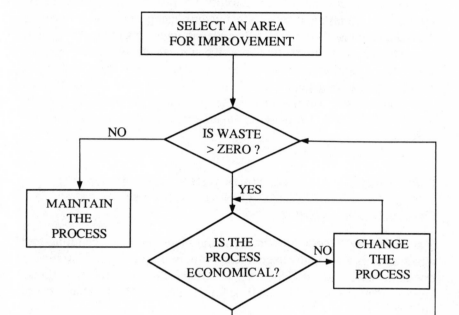

*BILL SANDRAS

For one company, it may be the materials area, focusing on inventory, sourcing, stockkeeping, and paperwork. For another, it's production, where the focus is on setup reduction, equipment layout and configuration, and mixed-model loading. It may be in accounting, focusing on standards, payment frequency, work-in-process accounting, labor accounting, and performance measures. Almost every company is affected in one way or another in the personnel area, whether it's in employee stability issues, flexibility between workers and job categories, communications, or responsibility for problem resolution. Engineering can be affected by product requirements, flow requirements, changing responsibilities in design and manufacturing engineering, and design teams. In the marketing area, it can mean changes to the way we forecast, sell, and market our products. Systems, processes, and facilities may need to change, especially in the areas of work orders, factory layout, and traffic. No matter what the area though, the Just-in-Time/Total Quality Control process is capable of providing tremendous results.

PART THREE—MANUFACTURING RESOURCE PLANNING AND JUST-IN-TIME/TOTAL QUALITY CONTROL OPERATING TOGETHER

Manufacturing Resource Planning is the planning and scheduling tool which enables a company to get the maximum performance from *today's* operating environment. However, it doesn't challenge the way things are—if there is a 52-week lead time for an item, MRP II just plans based on that.

Just-in-Time/Total Quality Control forces constant improvements to today's environment, so that tomorrow's environment is better. It's like the conscience of the system, always asking "can we do this better, more effectively, more economically, etc."

Manufacturing Resource Planning then plans and controls that new environment to enable a company to maximize performance out of *tomorrow's* improved environment. And so on, and so on. MRP II provides the tool, and JIT provides the process to make the environment better so the tool is more effective.

CONCLUSION

Faster, more reliable, and more economical responses to changes are the major advantage of having MRP II and JIT in place. Changes in the marketplace are always occurring, and some represent important opportunities and require quick analysis to determine the best way to take advantage of them. In addition, unavoidable problems are also occurring. Being unaware that the plans cannot be carried out only leads to bigger problems—not only do you lose valuable time before you recognize the situation, but during this period of time you are likely to compound the problem. "Bad news early is better than bad news late" is a better approach. It permits you to react to the problem before it's a crisis, rather than after.

Companies that operate MRP II and Just-in-Time/Total Quality Control in an outstanding manner are equipped to manage change—they have controls enabling them to lay out alternative plans, predict the consequences of each, select the best one, make it happen, and then continuously challenge each of the essential steps to make them even more effective. In short, they have greater control in running the business and continuously improving the business.

Thanks go to Walt Goddard, Bill Sandras, and Tom Wallace for the information contained in this appendix.

Appendix C

Sources for Additional Information

Preparing yourself to implement a Class A MRP II system requires careful study of a huge amount of information, far more than could be included in this or any other book. The Oliver Wight Companies can provide further assistance in getting ready, including books on the subject, live education, and reviews of commercially available software packages.

OLIVER WIGHT PUBLICATIONS, INC.

Oliver Wight Publications, Inc. was created in 1981 to publish books on planning and scheduling, written by leading educators and consultants in the field.

A complete library of books on Manufacturing Resource Planning, Just-in-Time, and Distribution Resource Planning are available.

For more information, or to order publications, contact:

Oliver Wight Publications, Inc.
5 Oliver Wight Drive
Essex Junction, VT 05452
800-343-0625 or 802-878-8161

Index